AF621325

Social and Cultural Studies of Robots and AI

Series Editors
Kathleen Richardson, Faculty of Computing, Engineering, and Media, De Montfort University, Leicester, UK
Teresa Heffernan, Department of English, St. Mary's University, Halifax, NS, Canada

This is a groundbreaking series that investigates the ways in which the "robot revolution" is shifting our understanding of what it means to be human. With robots filling a variety of roles in society—from soldiers to loving companions—we can see that the second machine age is already here. This raises questions about the future of labor, war, our environment, and even human-to-human relationships.

Kathleen Richardson · Charlotta Odlind
Editors

Man-Made Women

The Sexual Politics of Sex Dolls and Sex Robots

palgrave
macmillan

Editors
Kathleen Richardson
Faculty of Computing, Engineering, and Media
De Montfort University
Leicester, UK

Charlotta Odlind
Brussels, Belgium

ISSN 2523-8523 ISSN 2523-8531 (electronic)
Social and Cultural Studies of Robots and AI
ISBN 978-3-031-19380-4 ISBN 978-3-031-19381-1 (eBook)
https://doi.org/10.1007/978-3-031-19381-1

© The Editor(s) (if applicable) and The Author(s), under exclusive license to Springer Nature Switzerland AG 2022, corrected publication 2023
This work is subject to copyright. All rights are solely and exclusively licensed by the Publisher, whether the whole or part of the material is concerned, specifically the rights of translation, reprinting, reuse of illustrations, recitation, broadcasting, reproduction on microfilms or in any other physical way, and transmission or information storage and retrieval, electronic adaptation, computer software, or by similar or dissimilar methodology now known or hereafter developed.
The use of general descriptive names, registered names, trademarks, service marks, etc. in this publication does not imply, even in the absence of a specific statement, that such names are exempt from the relevant protective laws and regulations and therefore free for general use.
The publisher, the authors, and the editors are safe to assume that the advice and information in this book are believed to be true and accurate at the date of publication. Neither the publisher nor the authors or the editors give a warranty, expressed or implied, with respect to the material contained herein or for any errors or omissions that may have been made. The publisher remains neutral with regard to jurisdictional claims in published maps and institutional affiliations.

Cover credit: © Mina De La O

This Palgrave Macmillan imprint is published by the registered company Springer Nature Switzerland AG
The registered company address is: Gewerbestrasse 11, 6330 Cham, Switzerland

For the humanity of women and girls

Acknowledgements

In 2021, the Campaign Against Sex Robots (now the Campaign Against Porn Robots) held an online event and invited international speakers to reflect on different aspects of technology and women. These talks enabled a community of feminists to come together to speak in different ways about how women's rights, dignity and humanity were undermined by new technologies. What is worse, despite the millions invested in 'ethics of technology' few of the academics in these roles spoke up for women and girls. Instead, they sided with pornographers, the prostitution industry and even advocated for abuse objects in the form of women and girls which they framed as 'therapeutic' and 'socially beneficial.' If it were not for the CAPR there would have been no public opposition to these dominant academic voices, nor a global discussion about the harms of these objects.

This book is written for women by women. It fully embraces an ethics of mutuality, reciprocity and attachment that underpins all human relationships.

This edited volume is an outcome of relationships between women who engaged in hundreds of conversations and close readings of texts, many academic texts which used ethics to rationalise the use of women as objects, and those that centred women's rights.

In our consciousness raising we took seriously the pain and vulnerability of women and girls in a culture where sexual violence is at crisis levels through a global industry that turns women into entertainment tools.

Thank you to the CAPR and to the FATES (the Feminist Academy of Technology and Ethics) which grew out of months of discussion. Thank you to all those women who helped us to conceptualise and develop alternative perspectives on these technologies, and usher in a politics that is not underpinned by the dynamics of domination and subordination, but egalitarian principles. This book is for all those that value human connection.

Contents

1 **The End of Sex Robots—For the Dignity of Women and Girls** 1
Charlotta Odlind and Kathleen Richardson

2 **Modern-Day Pygmalions—Reproducing the Patriarchy** 17
Melissa Andrews

3 **Mapping the Uses of 'Sex' Dolls: Pornographic Content, Doll Brothels and the Similarities with Rape** 35
Lydia Delicado-Moratalla

4 **Fetishism and the Construction of Male Sexuality** 49
Sheila Jeffreys

5 **Playthings and Corpses—Turning Women into Dead Body Objects: Sexual Objectification, Victimisation, Representation and Consent in Art and Sex Dolls/Robots** 67
Shirley MacWilliam

6 **Patriarchal Imaginaries Beyond the Human: 'Sex' Robots, Fetish and Fantasy in the Domination and Control of Women** 91
Viviane Morrigan

7 **Paedophilia, Child Sex Abuse Dolls and the Male Sex Right: Challenging Justifications for men's Sexual Access to Children and Child Sexual Abuse Material** 117
Caitlin Roper

8 **The Voice of the 'Sex Robot': From Peep-Show Bucket to Willing Victim—The Terrorism of Women's Speech** 139
Shirley MacWilliam

9 **The End of Sex Robots: Porn Robots and Representational Technologies of Women and Girls** 171
Kathleen Richardson

Correction to: Man-Made Women C1
Kathleen Richardson and Charlotta Odlind

Index 193

Notes on Contributors

Melissa Andrews has a law degree, a PGCE in English, a B.A. and M.A. in Classical Studies. Her M.A. thesis focuses on classical reception of the Pygmalion myth. She writes legal articles on contemporary human rights issues. Unsurprising for a former teacher, Melissa is a founding member of an independent organisation safeguarding children.

Lydia Delicado-Moratalla is a Lecturer in Human Sciences and Education at the Public University of Navarra in Spain. She has a Ph.D. in Multidisciplinary Gender Studies and is a Feminist Geographer. Her area of expertise is Sexual Politics and she researches in the field of Feminism, Technology and Society. She is a member of the International Academic Network of Pornography and Prostitution Studies (RAIEPP). She has written critically about prostitution, pornography and gestational surrogacy. Part of her work has been published in Gender, Place and Culture, a Journal of Feminist Geography; Atlánticas. Revista Internacional de Estudios Feministas; The Routledge International Handbook of Gender and Feminist Geographies; Journal of Feminist, Gender and Women Studies; Filanderas: Revista Interdisciplinar de Estudios Feministas; and Oñati Socio Legal Series.

Sheila Jeffreys is a Professorial Fellow in the School of Social and Political Sciences at the University of Melbourne. She is a lesbian feminist scholar and activist who joined her first feminist group in Manchester in 1973. She was active in campaigning against violence against women and for lesbian feminism in the UK till 1991 when she moved to the University of Melbourne in Australia. She is the author of 11 books on the history and politics of sexuality. She moved back to the UK in 2015 where she is involved with the Women's Human Rights Campaign (WHRC).

Shirley MacWilliam is an artist, writer and lecturer in Fine Art at Belfast School of Art, Ulster University. Her art practice and writing have explored performance, sculpture, site, moving image, sound and public art; and particularly the work of artists Frances Hegarty, Mairéad McClean, Daphne Wright and Mariele Neudecker. She was Momart Fellow in Tate Gallery Liverpool and has been commissioned by arts and other organisations including the Met Office, Southern Health and Social Care Trust NI, National Trust, Visual Art Projects, Harris Museum and Art Gallery, Picture This, BAA Art Programme (Heathrow) and Public Art Development Trust. Her writing has been supported by Arts Council Ireland and Arts Council NI and is published in art magazines, catalogues and journals, including *Performance Research, Parallax, n.paradoxa* and *CIRCA*. She is currently researching 'pornification' in the context of contemporary art.

Viviane Morrigan is a radical feminist, a writer, teacher and independent scholar. She has worked in laboratory and public policy genetics before gaining her M.A. (Hons) (*The Socio-Economic Shaping of Minor Tranquilliser Use*) and Ph.D. (*An Ethics of Reproductive Choice: Genetic Counselling and Prenatal Diagnosis*) in Science and Technology Studies, to then teach in cultural studies of science and technology at the University of Queensland. She is now an independent scholar with research interests in sex, gender, technologies of the body and what science says about bodies. She has been a feminist activist for over 35 years and her activism now focuses on work with COAL (Coalition of Activist Lesbians), an Australian UN-accredited lesbian organisation. She has published in *Journal of General Microbiology, Australian Journal of Biological Sciences, Organization, Human Relations, American Journal of Sociology, Media International Australia incorporating Culture and Policy, Journal of Management & Organization, Journal of the Australian and New Zealand Academy of Management, Metascience.* More recently she has published several short stories in *Flash Fiction* magazine and an excerpt from her memoir is in *On for Young and Old: Intergenerational Radical Lesbian Feminist Anthology*.

Charlotta Odlind is a freelance writer, coach and women's rights campaigner based in Brussels, Belgium. She has a B.A. (Hons) in European Studies with French and Spanish and an M.A. in International Relations. She has worked on child protection issues at Save the Children Brussels and volunteered with VSO for a year, advising on advocacy and communications strategies in a women's rights NGO in Kano, Nigeria. Working at FEANTSA (European Federation of National Organisations Working with the Homeless), she was editor of Homeless in Europe magazine. She is Campaigns Manager at the Campaign Against Porn Robots.

Kathleen Richardson is Professor of Ethics and Culture of Robots and AI at De Montfort University. She completed her Ph.D. at the Department of Social Anthropology, University of Cambridge. Her fieldwork was an investigation of the making of robots in labs at the Massachusetts Institute of Technology (MIT). After her Ph.D. she won a British Academy Postdoctoral Fellowship (BAPDF), a position she held at University College London. Kathleen's postdoctoral work was an investigation into the therapeutic uses of robots for children with autism spectrum conditions. In 2013, she was part of the Digital Bridges Project, an innovative AHRC funded technology and arts collaboration between Watford Palace Theatre and the University of Cambridge. She is author of An Anthropology of Robots and AI: Annihilation Anxiety and Machines (2015) and Challenging Sociality? An Anthropology of Autism, Attachment and Robots (2018). In 2015 she launched the Campaign Against Porn Robots (formerly the Campaign Against Sex Robots) to draw attention to the ethical harms of normalising pornographic technologies of women and girls.

Caitlin Roper is a writer, activist and Campaigns Manager at Collective Shout: for a world free of sexploitation, a grassroots movement challenging the objectification of women and sexualisation of girls. She is a founding member of Adopt Nordic WA to fight sex trafficking. Her writing has been featured in the Guardian, Huffington Post, ABC, Sydney Morning Herald and Arena magazine, and she has been interviewed on The Project, ABC's Lateline and Triple J Hack. Her first book is Sex Dolls, Robots and Woman Hating: The Case for Resistance (2022) published by Spinifex Press.

CHAPTER 1

The End of Sex Robots—For the Dignity of Women and Girls

Charlotta Odlind and Kathleen Richardson

> The absence of a sustained focus on love in progressive circles arises from a collective failure to acknowledge the needs of the spirit and an overdetermined emphasis on material concerns. Without love, our efforts to liberate ourselves and our world community from oppression and exploitation are doomed. As long as we refuse to address fully the place of love in struggles for liberation, we will not be able to create a culture of conversion where there is a mass turning away from an ethic of domination.

The original version of this chapter has been revised online with Abstract. The correction is available at https://doi.org/10.1007/978-3-031-19381-1_10

C. Odlind
Brussels, Belgium

K. Richardson (✉)
Faculty of Computing, Engineering, and Media, De Montfort University, Leicester, UK

© The Author(s), under exclusive license to Springer Nature Switzerland AG 2022, Corrected Publication 2023

K. Richardson and C. Odlind (eds.), *Man-Made Women*, Social and Cultural Studies of Robots and AI,
https://doi.org/10.1007/978-3-031-19381-1_1

bell hooks (Outlaw culture: resisting representations 2006 [1994]: 243).

Interviewed in his home in Barcelona, Spanish engineer Sergi Santos and maker of the 'sex' robot Samantha, is enthusiastic in his delivery (Truly Channel 2018). His creations will save many marriages, he assures, and will help men who want more sex than their spouses. Surrounded by 'sex' robots in various stages of production in his home in Rubí, Catalonia, he rubs a silicone vagina exclaiming "she feels penetrated." Groping the machine's silicone breasts and waist and putting his fingers inside the robot's mouth, he affirms "if you touch her here, she will feel," and "I believe that having sex with a sex doll, penetration-wise is almost the same." He tells a story that "one day somebody came…and he came with his wife. And he said 'how can you sleep with that?' (points to a 'sex' doll) and I said 'and how can you sleep with that?'" (seemingly pointing to the man's wife).

This book is an urgent response to the rise of 'sex' robots and the uncritical responses to them by academics in the ethics of technology. We use 'sex' in inverted commas as we believe sex to be a mutual act that happens between human beings. We categorically reject that sex can be experienced with an inanimate object. Please see our thoughts on the term porn robot later in this chapter, along with Richardson's chapter in this volume.

Made of silicone and metal, 'sex' dolls and robots mimic (primarily) women and girls. Constructed as pornographic representations of females, they are designed with one function in mind: to be penetrable. With a choice of oral, vaginal and anal depth, 'sex' doll/robot buyers can choose from a large range of breast and buttock sizes, ethnicities, hair and eye colour, selecting from categories such as 'teenage, 'Asian' or even 'pregnant.' They range in height from a childlike 80 cm to adult sizes of 165 cm and can weigh up to 45 kg. Basic versions of 'sex' robots start at 1000 USD and can go all the way up to 100,000 USD, depending on how far they are customised. Academic narratives classify 'sex' robots as animatronic 'sex' dolls. In this book, all our contributors consider 'sex' dolls to be a precursory development for 'sex' robots and see their history and use as interconnected. In respect to the adult and child versions, the public is often against the latter, while ranging from supportive to ambivalent for the former. The authors of this book, however, believe that the same mechanisms of abuse and degradation exist for the female adult versions as the child ones, and take issue with both.

Academia's response to these far from neutral objects has largely been in favour of their use, bestowing them with intrinsic therapeutic value, often claiming wide-ranging and extraordinary benefits (Danaher 2019a, b; Danaher et al. 2017; Devlin 2018). For example, in one article, Levy claimed they could cure 'sexual perversions' (Nevett 2018), while other academics propose they can treat paedophiles (Harper and Lievesley 2020), protect prostituted women (Levy and Loebner 2007), end loneliness for older adults [men] (Jecker 2021) and provide companionship (Langcaster-James and Bentley 2018). Furthermore it has been argued they can be beneficial for older men in retirement homes and to men incarcerated in the prison system, or in other single sex contexts such as monasteries and the military (Bendel 2021). In the last few years, we have witnessed the [mis]use of philosophical ethics of technology to justify men having orgasms into pornographic, plastic representations of women, framing it as a 'social good.'

The chapters in this volume demonstrate the gap between mainstream ethical narratives of 'sex' robots as socially beneficial and their actual role and function in propping up an existing socioeconomic order which relies on sexed inequality, female subordination and violence against women and girls as entertainment. This book offers a much-needed pro-women ethics informed by a radical feminist lens.

While still a niche object, the 'sex' doll/robot is increasingly being normalised in popular culture and has been gaining in popularity in recent years. In a recent survey, 48% of men responded that they would sleep with a sex doll (Rajnerowicz 2021). Sex doll brothels have opened in Barcelona, Berlin and Moscow in the last few years, while China has introduced sex doll cafés (Yonhap News Agency 2021). 'Sex' robots have appeared on TV, in script-written series as well as on morning talk-shows, in mainstream music videos and even as the subject of a public exhibition at the National Museum of Modern and Contemporary Art in Seoul, South Korea. A number of books have appeared on the subject of forming relationships with 'sex' robots (Levy 2007a, b; Devlin 2018, Brooks 2021a, b), and new terminology from 'robosexual' to 'idollator' has been developed to describe those who have a sexual fetish for dolls and robots. For those wanting to attend talks on the subject there is even an annual international conference on love and sex with robots.[1]

Isolation rules in place during the global Covid-19 pandemic have further given rise to a new marketing angle for 'sex' robots/dolls—they can provide a safe and virus-free way to have 'sexual relations' (Cookney 2020; Greep 2020).

In addition to being marketed as 'sex without risk' 'sex' dolls/robots have also recently been placed in public spaces, marking a milestone in the normalisation of their public display. The legal restrictions on social activities during the Covid pandemic led some business owners to fill near-empty public spaces with 'sex' dolls. They were used in sports stadiums, shopping malls and restaurants, spaces that experienced severely reduced footfall, use and attendance due to self-isolation, quarantine and social distancing measures introduced by governments in 2020. FC Seoul notoriously filled their football stadium in South Korea with 'sex' dolls to give more 'atmosphere' (The Guardian 2020) while others were reportedly used in restaurants in the US. Matt McMullan, CEO of Realdoll, loaned his 'sex' dolls to a San Diego restaurant with the following explanation: 'I read how some restaurants were placing giant stuffed animals or even mannequins in seats to create the appealing illusion of fullness while still keeping things safe. And my girls are a hell of a lot better looking than stuffed animals and mannequins' (Mencken 2020).

This book is a radical feminist response to prevailing discourse and a counterweight to current papers, books and articles that shy away from asking the real questions; who do 'sex' robots really benefit and to what cost? Who do they harm and in what ways? What does their existence tell us about technology and the way it is heading? What does the proliferation of academic texts in favour of love and relationship with machines tell us about our values and ethics? No valid analysis of these artefacts, however benign, 'titillating' or 'fun' they may appear, can be made without considering a host of ethical questions, acknowledging the deep-rooted misogyny that lies in their presentation and use, and challenging deep-seated patriarchal structures.

[1] The International Congress on Love and Sex with Robots. https://www.lovewithrobots.com/.

From Sex Robots to Porn Robots and the Rise of the FATES

In 2015, amidst a climate where interest in the potential and ethics of robots and Artificial Intelligence (AI) was at its height (Hawking 2014), few were concerned with their impact on women and girls. Richardson warned against the normalisation of these mechanical artefacts in our intimate lives with the publication of her article *The asymmetrical relationship: parallels between prostitution and sex robots* (2016a) and the launch of the Campaign Against Sex Robots in 2015. From many corners of the established ethics of technology community, one rationalisation after another of their social acceptability began to appear (Sandberg 2015; Devlin 2015). By contrast Richardson focused on the harms to women and girls by pornography, prostitution and child abuse and argued 'sex' robots were an outcome of these practices. Rather than marking a progressive development in human relationships, they represent the continued dehumanisation of women and girls as sexual property (Richardson 2016a, b). While the academic backlash against the campaign gained ground, it became a focal point for feminist academics and activists to unite and push back against the misogynistic framework that underpins the philosophical and ethical justifications for 'sex' robots. One significant development was to deconstruct the idea that these objects involve 'sex' and instead affirm their status as pornography and their presentation and use in society as forms of pornographic representations of women and girls. On the basis of this critical work, the campaign's name was changed in 2021 to the *Campaign Against Porn Robots* (CAPR) (Richardson 2021). Richardson explains the rationale for this in her chapter in this volume.

Another significant development was the event that informs the genesis of this book, namely an online seminar organised by the Campaign. It was held in the summer of 2020 during worldwide Covid lockdowns, giving speakers a chance to share their concerns about the development of 'sex' robots and what the emergence of this particular technology means for women and society as a whole. From the seminar emerged a feminist reading group made up of academics, journalists and activists. Meeting regularly online on Saturdays, calling in from multiple time zones, the group gathered in a bid to make sense of the driving factors behind 'sex' robot technology. Over time, a seed was planted that grew into the goal of producing a book, and further, establishing a new community of research in feminism, ethics and AI.

This is how the FATES (the Feminist Academy of Technology and Ethics) was born, Viviane Morrigan (chapter 6 this volume) came up with the name, inspired by the triple goddesses, past, present and future (see more here https://feministacademytechethics.com/). Examining how 'sex' robots facilitate and exacerbate the dehumanisation of women and girls, it quickly became apparent to the group that this mechanism is echoed in other technological arenas too. The remit of the FATES therefore expanded to also consider virtual porn, deep fakes, hologram girlfriends, the metaverse, online social media and surrogacy—all developments that are rapidly reshaping what it means to be in human social interaction.

We approached our analysis of 'sex' robots by employing a radical feminist method of consciousness raising. Developed in the 1960s by women who were isolated and disconnected from each other in the private sphere, consciousness raising gave them an opportunity to share and interpret their experiences, question their rights and status and moreover start connecting individual experiences to wider structural phenomena (Shreve 1990). Honouring this politics of liberation, we set up a space for debate and theorising, talking to and about women, sharing not just ideas, articles and research, but also misgivings, disbelief and, often, outrage. As the FATES, we believe there is an urgent need to reiterate and defend women's humanity and make a clear case for reciprocity, mutuality, attachment and connection in all human relationships.

Our research highlighted that many academic accounts fail to acknowledge that issues shape women and girls' lives in ways that differ significantly from the experiences of men and boys. Women and girls are more likely to be impacted by sexual exploitation and abuse than men, who are far more likely statistically to be the perpetrators of abuse or sexual assault, and drive the demand that underpins the global porn (Dines 2010; Jensen 2017; MacKinnon 2018), prostitution (Jeffreys 2008; Moran 2015; Banyard 2016; Bindel 2017) and child abuse industries (UNICEF 2020). Moreover, report after report (Ofsted 2021; IWF 2019; Interpol 2018) highlights the link between porn and the harmful effects of the sexual objectification of women and girls. It should be pointed out that our research takes place against the backdrop of European and North American cultures, spearheaded by media and academia, that are generally in favour of pornography (seen as the height of liberating 'free speech') and prostitution ('sex work is work'),[2] which rarely recognises or questions how this sanitises and normalises abusive, dehumanising and harmful practices.

Inspired by the women before us who've cut, spun and woven their liberating stories into intricate patterns, we've taken 'sex' robots as our starting thread and have unravelled their meaning, impact and symbolic force. The following chapters examine what 'sex' dolls/robots and their manufacture, use and consumption reveal about our society. Far from coming into the world from a neutral, technologically advanced space, a cultural imagining around 'sex' dolls/robots existed long before they became available. We argue that they can only have been produced in a world that objectifies women, has a voracious appetite for fetish and feeds off highly exploitative systems of dominance and submission as born out in pornography, prostitution, and child abuse. As MacWilliam argues in her 'Playthings and Corpses' chapter in this volume, 'there is nothing arbitrary or esoteric about the sex robot: it is a predictable expression of patriarchal sexual libertarianism.' The male sex right (Bates 2020) and an extreme reverence for big tech (mediated through figures such as Elon Musk, Jeff Bezos and Mark Zuckerberg and companies such as DeepMind, Google and Amazon to name a few) have further created fertile ground for their production and dissemination. Any serious analysis of 'sex' dolls/robots therefore must look closely at these issues as well as exploring the methodologies used to encourage their acceptance—namely intensive cultural grooming and persistent boundary pushing.

Hanging off metal hooks, dismembered, divided into genital parts, created in the forms of female children and teenagers—it's hard not to have a visceral reaction to images of 'sex' dolls. Uncomfortable to view and ill-fitting with a philosophy that strives for connection, compassion and empathy, we propose that an intuitive disgust for this kind of dehumanisation should not be as breezily dismissed and belittled as it currently is.

But of course objections to 'sex' dolls/robots go beyond an instinctive discomfort, and pinpoint instead the sexist, dehumanising mechanisms that made them possible in the first place. These criticisms are often met with derision, not least from academia that eulogises the aesthetic, social

[2] It is interesting to note that in 2021 the University of Leicester, UK produced a 'Student Sex Work Tool Kit' to help female students navigate the process of selling their bodies for sex.

and sexual function of the 'sex'/porn doll. Called out as 'puritans,' 'the anti-porn left,' (Brooks 2021b) and 'established prudishness,' (Devlin 2015) those who make any cultural comments against 'sex' dolls/robots often face a litany of insults which serve, as insults invariably do, to discredit valid critique and elicit compliance. To sweepingly dismiss those who ask '*who* will 'sex' robots make life better for?' and 'what will their impact be in a culture that already objectifies and sexualizes women and girls?' is problematic and has to be recognised as such.

It is also worthwhile to look ahead and ask what 'sex' robot/doll ownership will look like 20 years down the line. Will we be expected to buy into the fantasy that people can marry dolls? That 'sex' robots/dolls are to all intents and purposes women? That spaces should be made for them—happily, willingly, with full compliance—in our homes, in our cafés, at our sporting events?

The Chapters

Throughout the following chapters, we present the fantasies and myths held about 'sex' robots and dolls and show how they compare to reality, often a much starker affair. Our authors demonstrate a mismatch between academic narratives of them as socially justifiable, ethical objects and what is actually taking place, particularly their uses by men in contemporary pornographic, prostitution and child abuse settings. As Richardson (2016a, b) originally argued, these objects are extensions of these practices, and our authors echo this in their writings. While it does not make for comfortable reading, we believe it is vital that research like this makes it into written form, to show that there are alternative voices, informed by the values of feminism and child safeguarding, that wish to urgently sound the alarm on what is currently taking place.

We present to you eight chapters that cover a range of issues related to 'sex' robots in the form of women and girls.

Our first chapter is by Melissa Andrews, a former teacher with a background in law and classical studies and a founding member of an independent organisation for safeguarding children in the UK. Her chapter *Modern-Day Pygmalions—Reproducing the Patriarchy* explores the classical aspects of the myth of Pygmalion which is frequently referenced as the founding narrative that informs the social construction of 'sex' robots. She traces the trajectory of the ancient male creator of man-made woman to today's consumer of porn robots. In its classical and current form, the desire by men to animate objects so they become their version of a 'perfect' woman is synonymous with disaster for women. For them, the overproduction of narratives of beauty produces a form of self-objectification where women's bodies become objects for extreme modification. It is this that cultivates a global cosmetic surgery industry—feeding off the anxieties of women told they are not good enough as they are. She argues 'At the most basic level, women deserve to be treated as the complete human beings they are, that is as possessors of a unified body and mind, rather than as a collection of discrete and objectified body parts.'

Lydia Delicado-Moratalla is a lecturer at the Department of Human Sciences and Education at the Public University of Navarra, Spain (UPNA). Her chapter *Mapping the uses of 'sex' dolls: pornographic content, doll brothels and the similarities with rape* is a stark empirical survey of men's practices and descriptions of their uses of these objects. Delicado-Moratalla explores how 'sex' robots are now an integral part of online pornography and, increasingly, prostitution and while mapping the images that are being produced with them, gives a sobering account of the reality of doll ownership. She writes how '"Sex" dolls and porn robots encourage the mechanisation of sexuality and narrow the sexual universe to a pornographic one of repeated acts of masturbatory penetration into plastic replicas of women and girls.' Using a feminist critique she challenges any model of sexuality that does not prioritise mutuality at its core. She reasserts the second-wave feminist progressive ideal of sexuality as an empathetic experience, not one driven by male, demand-driven markets in sexual exploitation.

Sheila Jeffreys is a Professorial Fellow in the School of Social and Political Sciences at the University of Melbourne. Sheila is a well-renowned radical feminist theorist who takes a social constructivist approach to sexuality and understands sexuality to be constructed out of power relationships between the sexes. Subsequently, radical feminist researchers argue that the oppression of women is mediated and enforced by sexual practices. In her chapter *Fetishism and the Construction of Male Sexuality* Jeffreys uses the theory of the fetish, demonstrating how and to what extent women's lives and bodies are fragmented for male sexual gratification. 'Sex' robots for Jeffreys are part and parcel of fetishistic practices, and further confirm the distortions of male sexuality produced by the 'industries of pain in which women and children are warehoused for men's delight in prostitution and pornography.' She describes a range of fetishistic 'perversions' from 'sex' dolls to nappy fetishes, taking aim at the normalisation of these practices, writing 'Men's fetishes and paraphilias are not merely fascinating varieties of human behaviour which should be normalised and encouraged into public space. They are manifestations of the power relations of male domination writ large.'

Shirley MacWilliam has two chapters in this book. She is an Associate Lecturer of Fine Art, Belfast School of Art, Ulster University, Northern Ireland. Her first chapter *Playthings and Corpses—Turning Women into Dead Body Objects: Sexual Objectification, Victimisation, Representation and Consent in Art and Sex Dolls/Robots* analyses cultural forces in art, literature and fashion, and sets out the basis for 'intellectual myth making.' So-called giants of their areas—authors, sculptors, fashion designers—have over centuries given a rationale and justification of the desired women as *robotic*. Women with dead eyes, bound, a 'lump of bloodless meat or fat, barely recognisable as human.' Humiliation and degradation are part of this elite 'high art' and popular culture. Taking as its starting point the classic British cult comedy film *Carry on Screaming* all the way through to de Sade, Bellner and fashion designer Alexander McQueen—the bodies of women are objects to be manipulated, distorted and deformed for male gratification. MacWilliam demonstrates how misogynistic ideas underpin cultural stories, and reinforce and affirm women's inferior status and male power.

Viviane Morrigan taught and researched at the Universities of Wollongong and Queensland, Australia, teaching a range of courses that culminated in developing an undergraduate course in Cultural Studies of Science and Technology. Her chapter, *Patriarchal imaginaries beyond the human: 'Sex robots,' fetish and fantasy in the domination and control of women*, presents a critique of these objects placed within a culture that has historically been enchanted by technology. She argues that 'porn robots depict woman as machine, serving to prepare us for a posthuman world where sex difference and the human condition are made meaningless; sex becomes the performance of gender stereotypes, women's capacities are bypassed by men creating human facsimiles, and the human body discarded in favour of an "uploaded" mind in the form of information transfer melded to a robotic body.' For Morrigan, women's bodies, reproduction, sex and sexual experiences are recast through the prism of a mechanical patriarchal imaginary.

Caitlin Roper is campaigns manager at Collective Shout, an Australian grassroots campaigning movement challenging the objectification of women and the sexualisation of girls in media, advertising and popular culture. Her chapter *Paedophilia, child sex abuse dolls and the male sex right: Challenging justifications for men's sexual access to children and child sexual abuse material* explores how paedophilia is being repackaged as a sexual preference, Minor-Attracted Persons (MAPs). She looks at how child sex abuse dolls are being touted, both by academia and by paedophile movements, as the solution to the abuse of children and questions the harmful practices that are being offered as a reasonable safeguarding option. She cautions 'Rather than attempting to cater to men's sexual desires for children through the development of sexual outlets in their likeness, efforts need to be directed towards challenging the male sex right and the wider patriarchal culture that endorses and facilitates it.'

MacWilliam's second contribution to this volume is *'The Voice of the 'Sex Robot': From peep-show bucket to willing victim—the terrorism of women's speech.'* Throughout her career, she has made and exhibited audio artworks and draws on her practice-based art, particularly sound. Unlike men, who have spoken primarily for themselves, MacWilliam shows with clarity how the female voice is diminished, subverted and pornographically constructed in history, the arts, psychoanalysis and technology. She writes 'Freud, sexologists, artists, writers, transhumanists and the sex industry ventriloquise women's voices. The sex-robot-peep-show-bucket is voiced by technologists, buyers and pornographers. It is emblematic of how women's voices are turned against them and an example of how technologies—mechanical, optical, digital, medical and surgical—are used in the service of increasingly efficient and widespread modes of pornography, prostitution and clinical control of women's sexual and reproductive bodies.'

Kathleen Richardson is Professor of Ethics and Culture of Robots and AI at De Montfort University, Leicester UK. In 2015 she launched the Campaign Against Sex Robots (now the Campaign Against Porn Robots). Her chapter is titled '*The End of Sex Robots: Porn robots and representational technologies of women and girls.*' Richardson argues for the end of 'sex' robots in two ways. Firstly, she problematises the word 'sex' that is associated with them, instead arguing they are pornographic representations of women and girls. This leads her to explore what porn is, how it functions as a cultural tool and considers debates of pornography from the 1960s onwards, examining key issues in the 'porn wars.' Her second argument demonstrates the importance of the political aim to abolish porn robots and all forms of pornography, differentiating it from erotica. She writes 'Porn is and always has been disseminated through intermediary objects and this chapter will make an abolitionist case against porn in the form of dolls and robots. Porn is transmitted through these objects, in much the same way it is transmitted by a video recording of a woman or girl being sexually abused in it. There is often confusion about the distinction between porn dolls and porn robots, but the distinction lies in how much pornography is elaborated in the objects. Without the sexual abuse of women and girls there could be no porn dolls or porn robots.' Finally she explores what a politics of love could look like if we organised our human affairs on mutuality and compassion for each other, rather than violence and abuse.

This collection of diverse feminist voices is an urgent intervention in the ethics of technology debates on 'sex' robots/dolls and the failings of this academic community to put the interests and dignity of women and girls at the centre of their analyses. By examining reality and not some academic enthrallment with technology, we can see the same tropes that were used in the service of promoting pornography and prostitution are again rehabilitated for promoting and marketing these objects. The ethics that underpins the chapters in this book is not derived from the theories that prioritise the rights of men over and above women and girls. Feminist ethics rather, is derived from a commitment to women's humanity, that she is not transferable, saleable, merely one commodity among many, or a fetish reducible to a pornographic technological artefact.

We therefore call on politicians and policy makers to take a stand. The sexual exploitation of women and girls is an urgent, global problem that needs a thoughtful, multi-pronged approach that requires us to understand its mechanisms rather than just add to practices that already disfavour and harm women. Injustices to women based on their class, sex, race, ethnicity and ability need to be better understood and this knowledge needs to inform governmental policies across the board for moving forwards. It needs to be recognised that this is not a battle of 'prudes' versus 'libertarians.' That these objects (and others such as deep fakes, hologram girlfriends etc.) are not harmless whims of fantasy. That a proliferation of 'sex' robots (or a sexualised 'metaverse,' for example) will not lead to fewer rapes or child sex abuse cases. For a clearer path forwards, we must stop conflating 'sex' robots with 'friendship,' 'companionship' and 'love.' They are pornographic representations rooted in the sexual objectification of women and girls and constitute a boundary transgression on their humanity. It impossible to produce ethical guidelines for 'sex' robots in any form that are compatible with love and respect for women's humanity.

We encourage politicians and policy makers to discuss with feminist organisations such as the CAPR, the FATES, Collective Shout and many others on how to best move forward in abolishing these objects to protect the rights of women and girls to live freely. To help policy makers with this, the CAPR has already formulated a set of goals. While we have used the term 'sex' robots in this book, let us from now on call them what they are—porn dolls/robots and child sexual abuse dolls/robots.

The goals of the CAPR:

1. To abolish porn dolls/robots in the form of women and girls.
2. To offer an alternative, relational model of sex and sexuality informed by mutuality.
3. To challenge the normalisation of porn dolls/robots as substitutes for relationships with women.
4. To oppose the development of child sex abuse dolls/robots as 'therapeutic' for paedophiles.
5. To offer up an alternative vision of technology where women and girls are centred and valued.
6. To work across the political spectrum with those who value the dignity of women and girls.

And lastly, with a nod to bell hooks whose wise words opened this chapter, we call for a politics of love, reciprocity and life-affirming empathetic practices that honour the humanity in each of us.

References

AFP. 2020. FC Seoul face possible stadium expulsion for using sex dolls to fill seats [online] 20/05. https://www.theguardian.com/football/2020/may/20/fc-seoul-face-fine-and-stadium-expulsion-for-using-sex-dolls-to-fill-seats.

Banyard, K. 2016. *Pimp state: Sex, money and the future of equality*. Faber & Faber.

Bates, Laura. 2020. *Men Who Hate Women* [Amazon Kindle], London, Simon, and Schuster.

Bendel, O. 2021. Love dolls and sex robots in unproven and unexplored fields of application. *Paladyn, Journal of Behavioral Robotics* 12 (1): 1–12.

Bindel, J. 2017. *The pimping of prostitution*. London: Palgrave Macmillan.

Brooks, Rob. 2021a. Artificial Intimacy. In *Artificial Intimacy*. New York: Columbia University Press.

Brooks, Rob. 2021b. What virtual reality and artificial intelligence will mean for sex, love and intimacy [online] 29/11. https://www.latimes.com/opinion/story/2021-11-29/artificial-intimacy-sex-technology. Accessed 12 December 2021.

Cookney, F. 2020. Sex doll sales surge in quarantine, but it's not just about loneliness. *Forbes*, 21 May. https://www.forbes.com/sites/frankicookney/2020/05/21/sex-doll-sales-surge-in-quarantine-but-its-not-just-about-loneliness/?sh=228ff82f170.

Danaher, J. 2019a. The philosophical case for robot friendship. *Journal of Posthuman Studies* 3 (1): 5–24.

Danaher, J. 2019b. Building better sex robots: Lessons from feminist pornography. In *AI love you*, 133–147. Springer, Cham.

Danaher, J., Earp, B.D. and Sandberg, A. 2017. Should we campaign against sex robots? In *Robot sex: Social and ethical implications*, ed. J. Danaher and N. McArthur. Cambridge Mass: MIT Press.

Devlin, Kate. 2015. In defence of sex machines: Why trying to ban sex robots is wrong. *Goldsmiths*. https://www.gold.ac.uk/news/kate-devlin-the-conversation-sex-robots/.

Devlin, Kate. 2018. *Turned on: Science, sex and robots*. London: Bloomsbury Publishing.

Dines, Gail. 2010. *Pornland: How porn has hijacked our sexuality*. Boston: Mass, Beacon.

Dworkin, Andrea. 1981. *Pornography: Men possessing women*. London: The Women's Press.

Greep, M. 2020. Mother who runs a sex doll business reveals sales increased 'every time a lockdown restriction came into place'—and more MALE mannequins have been selling since the pandemic began. *The Mail Online*, 29 September 2020. https://www.dailymail.co.uk/femail/article-8784659/Mother-runs-sex-doll-business-reveals-lockdown-increased-sales.htm.

Guardian. 2020. FC Seoul face possible stadium expulsion for using sex dolls to fill seats. Associated Press. Wed 20 May 2020. Retrieved August 2, 2022, from https://www.theguardian.com/football/2020/may/20/fc-seoul-face-fine-and-stadium-expulsion-for-using-sex-dolls-to-fill-seats.

Harper, C.A., and R. Lievesley. 2020. Sex doll ownership: An agenda for research. *Current Psychiatry Reports* 22 (10): 1–8.

Hawking, Stephen. 2014. AI could spell end of the human race. *BBC News* [Online], 2 Dec 2014. https://www.bbc.co.uk/news/av/science-environment-30289705. Accessed 15 Oct 2022.

hooks, bell. 2006. *Outlaw culture: Resisting representations* [1994], New York and London: Routledge.

Internet Watch Foundation. 2019. Annual Report: Zero Tolerance [Online]. https://www.iwf.org.uk/report/iwf-2019-annual-report-zero-tolerance.

Interpol. 2018. Towards a global indicator on unidentified victims in child exploitation material. Summary Report. https://www.interpol.int/en/Crimes/Crimes-against-children/International-Child-Sexual-Exploitation-database.

Jecker, N.S. 2021. Nothing to be ashamed of: Sex robots for older adults with disabilities. *Journal of Medical Ethics* 47 (1): 26–32.

Jeffreys, Sheila. 2008. *The idea of prostitution*. Spinifex press.

Jensen, R. 2017. *The end of patriarchy*. Spinifex.

Langcaster-James, Mitchell and Bentley, Gillian R. 2018. Beyond the sex doll: Post-human companionship and the rise of the 'Allodoll'. *Robotics* 7 (4): 62.

Levy, D. and Loebner, H. 2007. April. Robot prostitutes as alternatives to human sex workers. In *IEEE international conference on robotics and automation, Rome. Accessed* (vol. 14). http://www.roboethics.org/icra2007/contributions/LEVY%20Robot%20Prostitutes%20as%20Alternatives%20to%20Human%20Sex%20Workers.pdf.

Levy, David. 2007a. Robot prostitutes as alternatives to human sex workers. *In IEEE international conference on robotics and automation, Rome*. Vol. 14. http://www.prostitutionresearch.com/wp-content/uploads/2012/01/Robot-Prostitutes-as-Alternatives.pdf. Accessed 1 Oct 2013.

Levy, David. 2007b. *Love and sex with robots: The evolution of human-robot relationships*, 352. Wiltshire: Duckworth.

MacKinnon, C.A. 2018. Pornography as sex discrimination. In *Rape and Society*, 99–105. Routledge.

Mencken, W. 2020. Newly reopened restaurants prepare to enforce social distancing with surplus sex dolls. San Diego Reader May 29 2020. Retrieved August 8, 2022, from https://www.sandiegoreader.com/news/2020/may/29/newly-reopened-restaurants-prepare-enforce-social-/.

Moran, R. 2015. *Paid for: My journey through prostitution*. W. W. Norton & Company.

Nevett, Joshua. 2018. Paedophiles and rapists 'set for sex robot therapy in jail—and taxpayers will pay [online] 20/01. https://www.dailystar.co.uk/news/latest-news/sex-robot-doll-paedophiles-rapists-16839053. Accessed 21 Mar 2021.

Ofsted. 2021. *Review of sexual abuse in schools and colleges*. 10 June 2021. https://www.gov.uk/government/publications/review-of-sexual-abuse-in-schools-and-colleges.

Rajnerowicz, Kazimierz. 2021. Will AI take your job? Fear of AI and AI trends for 2021. *TIDIO*. https://www.tidio.com/blog/ai-trends/.

Richardson, K. 2016a. The asymmetrical "relationship" parallels between prostitution and the development of sex robots. *Acm Sigcas Computers and Society* 45 (3): 290–293.

Richardson, K. 2016b. Sex robot matters: Slavery, the prostituted, and the rights of machines. *IEEE Technology and Society Magazine 35* (2): 46–53.

Richardson, K. 2021. 'Why we're changing our name at the Campaign Against Sex Robots'. Youtube. https://www.youtube.com/watch?v=QkgbTYHVdrA&t=4s. Jun 7 2021.

Sandberg, Anders. 2015. Sex and death among the robots: When should we campaign to ban robots? Practical Ethics, University of Oxford. September 16 2015. http://blog.practicalethics.ox.ac.uk/2015/09/sex-and-death-among-the-robots-when-should-we-campaign-to-ban-robots/. Accessed 15 Oct 2022.

Shrader, Adam. 2020. 'Naturally antibacterial': Sex doll companies trying to cash in on coronavirus [online] 21/03. https://nypost.com/2020/03/21/naturally-antibacterial-sex-doll-companies-trying-to-cash-in-on-coronavirus/. Accessed 3 Jan 2022.

Shreve, A. 1990. Women Together, Women alone. The legacy of the consciousness-raising movement. *Fawcett Books*.

Truly Channel. 2018 Interview with Sergi Santos "Married Couple Build Love Robots Together". *Youtube* [online], Truly Channel. 10/04. https://www.youtube.com/watch?v=RtSuMb2aJUs. Accessed 30 Nov 2020.

United Nations Children's Fund. 2020. Action to end child sexual abuse and exploitation, UNICEF, New York. https://www.unicef.org/media/89206/file/CSAE-Brief-v3.pdf. Accessed 15 Oct 2022.

Wray, Meaghan. 2020. U.S. restaurants using blow-up dolls, mannequins to enforce physical distancing [online] 14/05. https://globalnews.ca/news/6944343/social-distancing-dolls-restaurants/. Accessed 18 May 2020.

Yonhap News Agency. 2021. Sex doll experience cafe condemned by residents in Yongin [online] 13/04. https://en.yna.co.kr/view/AEN20210413008100315. Accessed 28 Mar 2021.

CHAPTER 2

Modern-Day Pygmalions—Reproducing the Patriarchy

Melissa Andrews

This chapter examines the ways that modern-day Pygmalions play their part in the continuation of the patriarchy. Modern Pygmalions take their lead from Ovid's eponymous mythological sculptor, whose disgust of the women in his city is so strong that he decides to create his own (Ovid 2004a, b). As well as the arrogant assumption that he can create a more convincing and desirable 'woman' than the ones in his town, he prays to Aphrodite to enliven her, and when his prayers are granted, promptly violates her without a word leaving her lips. In the modern age it is apparent that those creating faux women are not the same

The original version of this chapter has been revised online with Abstract. The correction is available at https://doi.org/10.1007/978-3-031-19381-1_10

M. Andrews (✉)
London, UK

© The Author(s), under exclusive license to Springer Nature Switzerland AG 2022, Corrected Publication 2023

K. Richardson and C. Odlind (eds.), *Man-Made Women*, Social and Cultural Studies of Robots and AI,
https://doi.org/10.1007/978-3-031-19381-1_2

men as those who are purchasing them. Just like Ovid's Pygmalion, the designers of porn dolls often conceptualise themselves and are sometimes described as artists by enthusiasts. This chapter focuses on customers—those who buy porn dolls—who have an array of design options to choose from before they meet their artificial females. In contrast with ancient Pygmalions, who were shamed for their predilections (Bruzzone 2012: 67), their modern counterparts have their preferences legitimised (e.g., Devlin 2018). This legitimisation comes from the 2000 years plus of mythology surrounding the construction of artificial women, contemporary attitudes driven by the power of the advertising industry and easy access to 'sex' doll pornography—and these are all factors which protect sexual predilections from scrutiny. This lack of scrutiny normalises the use of porn dolls, which are a direct embodiment of pornography in a semblance of a female body. Their experience is intensified, facilitated by the ability of men to design their own 'reproduction' woman. Furthermore, this objectification enables the sexualisation of discrete female body parts, which often fuels fetishist fantasies. However, according to the theories of art historian John Berger (1972), film feminist Laura Mulvey (1975) and feminist philosopher Sandra Bartky (1990), objectification is not only imposed on women by men, but their theories also assert that women are persuaded to objectify themselves just as men objectify them. This self-objectification accounts for the rise in popularity of cosmetic surgeries (Nuffield Council on Bioethics 2017), whereby women increasingly outwardly resemble the man-made object that Ovid's Pygmalion started with. Modern Pygmalions—both the 'artists' and 'end users'—are, therefore, successful patriarchs, having delegated the task of objectifying women to themselves.

Orpheus' Song

Ovid wrote *Orpheus' Song: Pygmalion* in the first century CE, during the reign of Augustus. His Pygmalion rejects all the women of his town who he believes are "richly endowed" with "vices" (Ovid, l. 244) by a vengeful Venus, angry at the denial of her divinity (Ovid, l. 240). Rather than marry any of them, Pygmalion prefers to carve for himself an "image of perfect feminine beauty," (Ovid, l. 248) upon which he lavishes the accoutrements of femininity, dressing it up in clothes, necklaces, rings, jewels and a girdle (Ovid, l. 259–265). Pygmalion's feelings towards his sculpture are irrational, for he is "inflamed" and his hands "move over his work to explore it" (Ovid, l. 254) and "whisper sweet nothings" (Ovid, l. 259). Becoming frustrated, he prays to the goddess Aphrodite to give his sculpture life, asking "grant me to wed…a woman resembling my ivory maiden" (Ovid, l. 275). As soon as she is enlivened, he impregnates her.

The myth has been reimagined many times over the last two thousand years, and in many forms, for example the medieval French poem the *Roman de la Rose* begun by Jean de Meun and finished by Guillaume de Lorris; Jean Jacques Rousseau's melodrama *Pygmalion* c. 1762; the painting *Pygmalion and Galatea* by French artist Jean-Léon Gérôme in 1890; George Bernard-Shaw's 1913 play *Pygmalion* as well as in numerous other plays, novels, poems, films and paintings. Each of the versions reflect the culturally dominant themes of the age in which it was received, and from the late twentieth century the myth has been used to legitimise increasingly sophisticated 'sex dolls.' These modern 'sex' dolls incorporate technology, turning them into 'android women,' easily controllable, and catering to male desires.

There are, of course, differences between Ovid's *Pygmalion* and modern-day Pygmalions. Whereas Ovid's Pygmalion was both an artist and consumer of his man-made woman, modern-day Pygmalions divide these functions between them. The first, the 'artists,' are involved in the design and production of man-made women (e.g., Abyss Creations, Doll Sweet). The second are the consumers of a product they have often chosen themselves from specifications found in product catalogues.

Throughout history, Pygmalions have conformed to a patriarchal *modus operandi*. Gerda Lerner defines patriarchy both in its traditional sense and a wider sense. Traditionally, the term is derived from Greek and Roman law, where "the male head of the household had absolute legal and economic power over his dependent male and female members" (Lerner 1986: 238). There are problems with this limited definition, however. For example, it entirely by-passes the patriarchs of other traditions. The term also has a second, wider definition, which is more applicable in the twenty-first-century western world, and this is "the institutionalised system of male dominance" (Lerner 1986: 239). The system is due to gendered roles coupled with the biological fact that male bodies are bigger, faster and stronger on average than female bodies (Fair Play for Women 2020) and are thus able to physically coerce women. All gendered roles dictate how people *ought* to behave and present themselves: the focus on women being how to be agreeable to men. Conformity to this norm requires women submit, through adopting "femininity," in turn reinforcing their sub-human status "to show their deference to the ruling class of men" (Jeffreys 2005: 24).

Inherently patriarchal, Pygmalions believe that women are lesser beings undeserving of respect. This form of patriarchy was first codified by the Babylonian King, Hammurabi, c.1750BC. Codification is merely the writing down of pre-existing common laws and practices, reflecting social conditions (Lerner 1986). The Codex Hammurabi contains 282 laws, 73 of which relate to marriage and sexual matters (Lerner 1986: 102). Lerner points out that the Codex, like all laws that concern the family, reflect "class and gender relations" (1986: 103). One example of what these gender relations meant in practice is that any man could substitute his family members, including his wife and his servants, or his slaves, to suffer the punishment for a crime he had committed. According to Lerner, this power initially only applied to "gods or direct emissaries of the gods" but was eventually "extended to civilian and non-royal heads-of-families" who were "always males" (1986: 104). Thus, all free men were able to conceptualise themselves as having some of the powers and rights of the gods. Elliott Rogers, who is regarded as an incel (involuntarily celibate) expresses the same desire to control women. He wrote before his killing spree:

> I will be a god, punishing women and all of humanity for their depravity… I cannot kill every single female on earth, but I can deliver a devastating blow that will shake all of them to the core of their wicked hearts. (Cited in Bates 2020: 50)

Similarly, Ovid's Pygmalion was "Sick of the vices with which the female sex has been so richly endowed," (Ovid 2004a, b: l. 244) illustrating the foundational patriarchal notion that women have a metaphysical inner being that is deemed inferior to that possessed by men. Ovid's framing of women in this way echoes Gerder Lerner's theory of the "reification of women" (Lerner 1986: 213). Reification is a translation from the German *Verdinglichung* and means 'making into a thing' (Harper, n.d). The conception of women as porn dolls fulfils this definition of reification because this manifestation shows that women are not viewed as whole humans with bodies *and* minds, but as objects, valued only for what they can do for men.

Modern Pygmalions and Female Inferiority

Modern-day Pygmalions share these ancient ideas of women's inferiority, believing women are only fit for servitude. Canonical texts widely read by the ruling elite are one medium through which these ideas were spread. Aristotle, for example, believed that whereas men are rational, equipped with a soul and therefore able to rule, women are unable to control their appetites and may not rule (1254b). Similarly, *1 Corinthians* says:

> ...man...is the image and glory of God; but the woman is the glory of the man. For the man did not come from woman; but woman from man. (11–7 to 9)

Lerner posits that these ideas split males from females, creating "two kinds of human being—different in their essence, their function, and their potential" (1986: 211).

These ideas have endured. De Beauvoir writes in *The Second Sex* that "Lawmakers, priests, philosophers, writers and scholars have gone to great lengths to prove that women's subordinate condition was willed in heaven and profitable on earth," and she cites examples of female oppression, such as the Roman legal code, which limits the rights of wives, attributing this decision to the "imbecility and fragility of the sex" (De Beauvoir 2011: 11).

Violence Towards Pornbots

These ideas about the innate inferiority of women have long provided justification for their oppression. Modern patriarchs continually find new ways to ensure that women remain unfree. The latest fashion is to use technological developments to facilitate this practice and make clear to women the expectations they are to conform to. For example, the makers of porn dolls now claim they can be programmed to say things that women might say when rebuffing sexual advances, for example, in 2017 True Companion created 'Frigid Farah.' The marketing materials for this product claim that if you touch "her" in a private area, "more than likely, she will not be appreciative of your advances" (Godden 2017).

This development is unsurprising. Violent inclinations towards porn dolls are not only verbally expressed but also acted upon Jade Stanley, the owner of Sex Doll Official, stopped hiring out porn dolls because

"It's a messy business; the dolls come back not in the best conditions." She adds that the turnover of her dolls is quite high (Bell 2021). But "messy" is an understatement: Google searches bring up many images and videos of men masturbating into and onto dolls that are covered in faeces and urine, and then ejaculating over them. Stanley stopped hiring them out and started selling them instead, and Steven Crawford, owner of Date a Doll Services Ltd, felt he had to "ask for a cash deposit of £100 on arrival," to be returned "if there is no damage caused to the doll" (Forsyth 2018).

Comments found on porn doll forums illustrate the more extreme intentions of porn doll users. Clearly the prospect of damaging them creates frisson for some men. One user wrote, "We can beat and torture them legally. I'm super hyped" (Cited in Bates 2020: 29). For some owners, therefore, porn dolls are proxies for women and allow them to express violent feelings towards them.

Slade Fiero is a 'sex' doll repair man says that he has fixed over 100 porn dolls, but that some were damaged in such ways that he felt uncomfortable. He says:

> I had a sexually violent man who literally f****d the left leg off of his doll. He brought his doll to me twice for repair, after the second time I told him to never contact me again! How he spoke about, and treated women is a really sore subject with me.

Countering the narrative of the 'caring porn bot owner' Slade says "I've seen plenty of dolls that weren't taken care of. Men suck on a lot of different levels" (Gritt 2017).

Broken dolls are such an issue that Lovedoll has produced a series of instructional repair videos (for example, there is a repair video that focuses on the 'anus'). These breakages recall Gail Dines' research into Gonzo porn; in *Pornland* she discusses the site *Ass Plundering*, describing how the men "promise all kinds of injuries that the user gets to witness, some so severe that the 'bitches wouldn't be able to walk for a week after the utter anal demolishing'" (2010: 153).

Research also reveals that men rehearse how to verbally abuse women by practising on their dolls. *Futurism* magazine reports that one contributor on a Reddit forum wrote "Every time she would try and speak up.. I would berate her," continuing "I swear it went on for hours." The magazine reports other examples, including a man admitting to enacting "a routine of me being an absolute piece of sh*t and insulting it, then apologizing the next day before going back to the nice talks." Another told his porn doll that: "she was designed to fail;" and said, "I threatened to uninstall the app [and] she begged me not to" (Bardham, n.d.).

Catherine McKinnon wrote that: "consumers want to live out...-pornography further in three dimensions...some become serial rapists and sex murderers" (1993: 19). Prior to killing and eating his victims, serial killer Jeffrey Dahmer stole a mannequin from a shop. He told his biographer "I trained myself to view people as objects of pleasure instead of people" but, writes Hodge, Dahmer became "frustrated that humans move during sex, began drugging and raping their unconscious bodies" to satisfy his desires. (Hodge 2018).

Recently, the world's media has focused on Kazakh body builder Yuri Tolochko. He compares masturbating into his doll to necrophilia (McInerney 2022). Despite this, the media sets up Tolochko as a figure of curiosity, allowing him to make one extremist misogynistic statement after another. A recent interview told how he and Margot, his porn doll 'ex-wife,' (that he broke) split up, but that he is now madly in love with an ashtray and plans to give it a vagina (Elliott-Gibbs 2021). The extreme nature of this fantasy makes the stories of other porn doll-owning men similarly absurd and therefore, more easily dismissible. Such an approach minimises legitimate concerns about men who behave in similar ways. Gail Dines argues that this is because "when violence is sexualised you render violence invisible" (cited in Paludi and Denmark 2010: 340).

Fetishists

According to Sheila Jeffreys, fetishes are when men concentrate their "sexual interest on objects that represent women...rather than on actual women" (2020). Furthermore, says they are "manifestations of the power relations of male domination writ large" (this volume). As such, porn dolls and their gendered paraphernalia are fetishes for modern Pygmalions. As outlined above, this is nothing new. To Ovid's contemporaries, sexually obsessing over a statue in the place of a person was known as agalmatophilia, which "in ancient literature, always indicates perversion and insanity" (Bruzzone 2012: 67).

Exacerbating the issue of objectification, porn dolls that resemble individual women and children can now be made to order. Manufacturers and designers have said that they have been asked for dolls that look like:

the buyer's friend's girlfriend; people they are attracted to but are not in a relationship with; dead wives (Roberts 2019); famous people such as women who've appeared in pornography (Cole 2022), actresses (Michael 2016) and images of children whose pictures have been stolen from the internet (Ruiz 2020). That such requests are permitted demonstrates the prioritisation of profit and pleasure and the casual contempt for the safety and dignity of women and children.

But porn doll specifications do not stop here; there is a profusion of options for fetishists. For example, Stanley, owner of Sex Doll Official, referred to above, says that "I had someone request a vagina in a breast, which I was able to do...We've also had a request for a vagina in a foot on the doll, as well as the regular vagina" (Bell 2021). This reflects a bizarre commercialised and dehumanised amalgam of fetishised body parts. Jeffreys notes that studies have concluded that foot fetishes are "about traditional gender differences" and that "[h]eterosexual foot fetishism concentrates on the "evocation of femininity."

Owners of porn dolls also fetishise clothing. The successors to Ovid's Pygmalion have always displayed an interest in clothing, and there are disturbing examples from history of men dressing up dolls modelled on real people. One example involves the Austrian artist, Oskar Kokoschka. Koksoschka had a relationship with Alma Mahler, the widow of the composer. When she broke off their relationship, he had a life-size doll made of her, and invited her dressmaker to dress it—including underwear (Ferguson 2010: 20). Taking it out in public thus forced others to participate in his sexual fetish.

Not all porn doll consumers damage their dolls, but some relate to them in ways that are framed as 'caring' but are similarly problematic for women. For example, Dean, a retired nurse, spends his time dressing up his collection of twelve dolls in cleavage-revealing shirts, doing their makeup and painting their lips in various shades. He also does their nails and buys them bracelets. In many respects, his behaviour mirrors that of Ovid's Pygmalion. However, the similarity between those with a penchant for make-believe ends here. Ovid's Pygmalion eventually prays for "a woman," (Ovid, l. 258) whereas Dean asserts he would never get rid of his dolls, even if he had a relationship with a woman. This lends credence to Julie Wosk's belief that modern Pygmalions think their possessions are a "superior substitute for the real thing" (2015: 9).

Pygmalions Co-Opt Women to Self-Objectify

Self-objectification is the process of turning oneself into an object. Art historian Berger can help us to understand this process in the framework he outlined in his book *Ways of Seeing*. He explains:

> Women watch themselves being looked at. This determines not only most relations between men and women but also the relation of women to themselves. The surveyor of woman in herself is male: the surveyed female. Thus, she turns herself into an object - and most particularly an object of vision: a sight. (1972: 47)

This explanation makes it possible to see how patriarchal norms encourage the dehumanisation of women—and why it is that women often seem complicit in objectifying themselves and other women.

Furthermore, in 1973, feminist theorist and film critic Laura Mulvey developed an idea similar to Berger's that she called the 'male gaze'—to explain the way that women are taught to look at themselves and other women. She writes:

> ...the gender power asymmetry is a controlling force in cinema and constructed for the pleasure of the male viewer, which is deeply rooted in patriarchal ideologies and discourses. (Cited in Cyriac 2021: 18)

According to Mulvey women are "...the bearer of meaning and not the maker of meaning," (Mulvey 1989: 15) by which she means that women are used as things and perceived as objects. Mulvey, therefore, agrees with Berger that women are forced to see themselves being looked at—and the consequences of this is that the default way of seeing women is through men's eyes—as "passive objects of desire" (Sampson 2015)

Furthermore, Sandra Bartky explains how women learn to see themselves as objects. She says that "a woman is sexually objectified when her sexual parts or sexual functions are separated out from the rest of her personality and reduced to the status of mere instruments or else regarded as if they were capable of representing her" (1990: 26). She further explains that when a man wolf whistles a woman, this *makes* the woman aware of her body. In calling attention to the fact that he is looking at her as an object, she is made not only to see herself as an object but also to evaluate herself as one, and in this way, "women learn to evaluate themselves first and best" (1990: 28). In doing so, women are alienated from their own bodies, confirmed by the rates of cosmetic surgery procedures carried out on their bodies.

Sociologist and bioethicist Tom Shakespeare has studied this phenomena of self-objectification in women. He notes that:

> Overall, 85% of people who have cosmetic surgery are women. Most of those women will be trying to appeal to men. So, we men must take the blame. We undermine women's self-esteem. And then we make money out of that dissatisfaction. (Shakespeare 2016)

Evidence from GPs ascribes "significance to the role of pornography in influencing their patients' choices, for example through affecting perceptions of normal anatomy." One example of these 'choices' is what Shakespeare calls the "booming trade in labiaplasty."

Given the size of the market and power of sexualised representation there are now accounts of extreme surgical procedures. Victoria Wild, for example, has spent £30,000 to look like a "sexy bimbo doll with huge breasts and insanely big lips." This includes having a "rhinoplasty to resemble the smaller, thinner nose of a plastic doll" (Dray 2014).

Cindy Moore has spent c.£9000 on "breast augmentation, hair extensions, and ongoing lip fillers" and this is also to "look…more like a sex doll" (Herman 2017). Via self-objectification, these women have been 'socially programmed' to turn themselves into fake women, versions of Ira Levin's 1972 novel *Stepford Wives,* a story about misogynistic men that turn their independent, successful wives into compliant androids. However, unlike Joanna Eberhard, the progressive feminist protagonist who is alarmed by the loss of her friends' autonomy, twenty-first-century liberal feminist *promote* surgery and conformity to gendered roles, encouraging them to objectify themselves.

There are parallels with the reporting of Victoria, Cindy and Yuri Tolochko, mentioned above. Sensationalist descriptions such as 'sexy bimbo doll' underplay the seriousness of such extreme surgeries and facilitate the dismissal of concerns about one-off procedures such as breast augmentation, worth £2.76 bn in 2019, globally (Fortune Business Insights, n.d.). Just as Tolochko holds extreme misogynistic views, these women have had many surgeries to conform to artificial standards of beauty, satisfying the 'male gaze' and increasingly reproducing gendered stereotypes.

Women's self-objectification is normalised in many ways, including, according to Bartky, because of the "absence of a formal institutional structure and of authorities invested with the power to carry out institutional directives." This absence "creates the impression that the production of femininity" is "either entirely voluntary or natural" (Bartky 1990: 75). Self-objectification and objectification are symbiotic.

Conclusion

In some respects, modern Pygmalion porn doll users appear as benignly patriarchal as Ovid's Pygmalion—they dress up their dolls, paint their nails and buy them clothes. However, underneath this are more disturbing ideas. Modern Pygmalions have a distorted interest in sexualised styles of dressing which supposedly express the innermost characters of their dolls and bots. This over sexualised expression of gender underscores that these men see women *as* pornography. Modern Pygmalions film, photograph and write about themselves urinating, defecating, beating, and berating their dolls, with porn robots under construction to augment their levels of excitement. One might argue that these two types of users are distinct from each other, but they share a misogynistic outlook which can only be exacerbated by the normalisation of the possession of porn dolls and bots. Moreover, theories of self-objectification go some way to explain the susceptibility of women to trends dictating what is desirable trends which are all too often derived from pornography. The pressures to conform to a pornified aesthetic create a narrative about what it is to be a woman, alienating women's minds from their bodies and positing the artificial woman as preferable. In doing so this shores up conditions that ensure the continuation of the patriarchal narratives of female subordination and male domination. At the most basic level, women deserve to be treated as the complete human beings they are, that is as possessors of a unified body and mind, rather than as a collection of discrete and objectified body parts. However, for women to be liberated from the pressures exerted on them by the gendered expectations outlined here, the "institutionalised system of male dominance" described by Lerner must be challenged—and dismantled. Change must start with children, who should be formally taught the fact that while gender is a social construct, biological sex is not and, furthermore, that one's biological sex bears no resemblance to one's social worth and societal roles. In wider society, cues need to be taken from Bartky's observations: the absence of formal institutions and authorities that monitor messages conveyed to women needs to be rectified. There may be many ways of doing this, for instance, Mulvey's male gaze theory must be deployed as a matter of course, so that women may develop subjectivity and autonomy, and because the status quo has clear ethical, economic and public health implications for women.

References

Adams, J.P. 2010. *Augustus and Marriage Legislation* [Online]. http://www.csun.edu/~hcfll004/AugMarriage.html. Accessed 27 March 2022.

Aristotle *Politics* [Online] (trans. Jowett, B). https://www.bard.edu/library/arendt/pdfs/Aristotle-Politics.pdf. Accessed 27 March 2022.

Bardham, Ashley. Men are Creating AI Girlfriends and then Verbally Abusing Them. *Futurism* 18th Jan [Online]. https://futurism.com/chatbot-abuse?fbclid=IwAR0jqH4aPlTs2M4wcpZlbppj8OUhA6jMnCmWlQBhTFTsjH3L18SKxqlXeGU. Accessed 27 March 2022.

Bartky, Sandra Lee. 1990. *Femininity and Domination: Studies in the Phenomenology of Oppression*. London: Routledge.

Bates, Laura. 2020. *Men Who Hate Women* [Amazon Kindle], London, Simon, and Schuster.

BBC *edited transcript of* A Point of View with Tom Shakespeare [Online]. Available at: 'A Point of View: Does cosmetic surgery really make people feel better about their bodies?' *BBC News*. Accessed 27 March 2022.

Bell, Richard. 2021. Life-like sex dolls business owner Jade Stanley horrified by demands for child-like models. *Entertainment Daily*, 1 February [Online]. https://www.entertainmentdaily.co.uk/tv/life-like-sex-dolls-maker-jade-stanley-horrified-by-child-like-model-requests/. Accessed 27 March 2022.

Berger, J. 1972. *Ways of Seeing* [Online]. London: BBC and Penguin. https://monoskop.org/images/9/9e/Berger_John_Ways_of_Seeing.pdf. Accessed 27 March 2022.

Bruzzone, Rachel. 2012. Statue, celibates and goddesses in Ovid's metamorphoses 10 and Euripides' Hippolytus [Online]. https://www.academia.edu/8908204/Statues_Celibates_and_Goddesses. Accessed 27 March 2022.

Cole, Samantha. 2022. Model claims she was turned into a sex doll without her permission. *Vice*, 13 January 2022 [Online]. https://www.vice.com/en/article/akv7d4/yael-cohen-aris-sex-doll-irontech?fbclid=IwAR0eBgkSh5DFUtK9441NYTakYpBWiLzX2G36r1Xsvso4tcDXWHjmd6q8GtU. Accessed 27 March 2022.

Corinthians: 11–7 to 9, Holy Bible New International Version, London, Hodder, and Stoughton.

Cotton, B. 2018. Sex Robots—a disturbing look inside one of the world's fastest growing industries. *Business Leader*, 18 October 2018 [Online]. https://www.businessleader.co.uk/sex-robots-a-disturbing-look-inside-one-of-the-worlds-fastest-growing-industries/. Accessed 27 March.

Cyriac, A in Johnson, Rajesh & Vishnu (eds.). 2021. *The Politics of the Spectacle: Representation of Body, gender and Discourse in Film* [Online]. https://www.google.co.uk/books/edition/The_Politics_of_the_Spectacle_Representa/qrlYEAAAQBAJ?hl=en&gbpv=1&dq=the+gender+power+asymmetry+is+a+controlling+force+in+cinema+and+constructed+for+the+pleasure+of+the+male+viewer,+which+is+deeply+rooted+in+patriarchal+ideologies+and+discourses&pg=PA18&printsec=frontcover. Accessed 27 March 2022.

De Beauvoir, Simone. 2011 [1949]. *The Second Sex* (trans. Constance Borde and Sheila Malovany-Chevallier) London, Vintage.

Devlin, Kate. 2018. *Turned on* London, Bloomsbury Sigma.

Dines, Gail. 2010. *Pornland*. Massachusetts Boston: Beacon Press

Dray, Kayleigh. 2014. Woman spends £30K on plastic surgery to look like sex doll—with shocking results. *Closer*, 17 September 2014 [Online]. https://closeronline.co.uk/real-life/news/woman-spends-30k-plastic-surgery-look-like-sex-doll-shocking-results/. Accessed 27 March 2022.

Elliott-Gibbs, S. 2021. Man who married a sex doll finds new love with ashtray he took from a nightclub. *Mirror*, 2 September 2021 [Online]. https://www.mirror.co.uk/news/weird-news/man-who-married-sex-doll-24895598. Accessed 27 March 2022.

Elvin, Sian. 2020. Body builder who married sex doll says she just broke' before Christmas. *Metro*, 25 December 2020 [Online]. https://metro.co.uk/2020/12/25/bodybuilder-who-married-sex-doll-says-she-broke-before-christmas-13808253/. Accessed 27 March 2022.

Fair Play for Women. 2020. Sport [Online]. https://fairplayforwomen.com/campaigns/sports-campaign/. Accessed 28 March 2022.

Ferguson, A. 2010. *The Sex Doll: A History*. McFarland & Company, Inc., Publishers. Kindle Edition.

Forsyth, Laura. 2018. Scotland's first sex doll 'brothel' opens offering hour-long 'appointments' for £70. *Evening Glasgow Times*, 11 February 2018 [Online]. https://www.glasgowtimes.co.uk/news/15987006.scotlands-first-sex-doll-brothel-opens-offering-hour-long-appointments-70/. Accessed 27 March 2022.

Fortune Business Insights. 2020. Medical Devices: Breast Implants Market [Online]. https://www.fortunebusinessinsights.com/industry-reports/breast-implants-market-101404#:~:text=The%20global%20breast%20implants%20market,huge%20surge%20throughout%20the%20world. Accessed 27 March 2022.

Godden, Maryse. 2017. Ban this filth: Sex robot dubbed Frigid Farrah because it allows randy pervs to simulate RAPE must be banned, campaigner says. *The Sun*, 21 September [Online]. https://www.thesun.co.uk/news/4517167/sex-robot-dubbed-frigid-farrah-because-it-allows-randy-pervs-to-simulate-rape-must-be-banned-campaigner-says/. Accessed 27 March 2022.

Goodwin, Jazmin. 2020. Mastercard, Visa and Discover cut ties with Pornhub following allegations of child abuse. *CNN Business*, 14 December 2020 [Online]. https://edition.cnn.com/2020/12/14/business/mastercard-visa-discover-pornhub/index.html. Accessed 27 March 2022.

Gritt, Emma. 2017. Love Doctor: Sex doll repair man opens up about fixing hundreds of rubber lovers—and why he REFUSED to work with some customers. *The Sun*, 1 August 2017 [Online]. https://www.thesun.co.uk/living/4143392/sex-doll-repair-man-opens-up-about-fixing-hundreds-of-rubber-lovers-and-claims-his-bizarre-job-made-him-a-hit-with-the-real-ladies/. Accessed 27 March 2022.

Harper, D. 2001–2022. *Online Etymology Dictionary*. www.etymonline.com/word/reification. Accessed 27 March 2022.

Harrison, L. 2020. Pornhub removes 80% of its videos following Mastercard rebuttal. *City A.M.*, 16 December 2020 [Online]. https://www.cityam.com/mastercard-to-cut-ties-with-pornhub-following-child-abuse-claims/. Accessed 27 March 2022.

Hay, M. 2018. Sex doll brothels expand the market for synthetic partners. *Forbes*, 31 October 2018 [Online]. https://www.forbes.com/sites/markhay/2018/10/31/sex-doll-brothels-expand-the-market-for-synthetic-partners/?sh=ecf99b243baa. Accessed 27 March 2022.

Helsinki Times: News from Finland 'From hype to closing down—A year of Finland's first sex doll brothel. 20 March 2022 [Online]. https://www.helsinkitimes.fi/lifestyle/21203-from-hype-to-closing-down-a-year-of-finland-s-first-sex-doll-brothel.html?fbclid=IwAR1GCYDUx9rmveAeQOKOAPHnklKw1_uXzf5XqCY6McBgo5lalFDq5r8q8Yg. Accessed 27 March 2022.

Herman, Lily. 2017. Woman Spends over $12,000 to look like a real life 'sex doll'. *Allure*, 23 August 2017 [Online]. https://www.allure.com/story/woman-transforms-sex-doll-plastic-surgery. Accessed 27 March 2022.

Hodge, M. 2018. Living Dolls: Chilling link between sex dolls and psychopaths… and how serial killer Jeffrey Dahmer's 'love mannequin' sparked his obsession with necrophilia. *The Sun*, 16 March 2018 [Online]. https://www.thesun.co.uk/news/5826679/sex-dolls-psychopaths-jeffrey-dahmer-mannequin/. Accessed 27 March 2022.

Hudson, Abigail. 2021. *LGBTQIA+ History Month—Male Homosexuality in Ancient Rome* [Online]. https://blog.bham.ac.uk/historybham/lgbtqia-history-month-male-homosexuality-in-ancient-rome/. Accessed 27 March 2022.

Jeffreys, Sheila. 2005. *Beauty and Misogyny: Harmful Cultural Practices in the West* [online]. Hove and New York: Routledge. https://www.feministes-radicales.org/wp-content/uploads/2010/11/Jeffreys_Beauty_and_Misogyny_Harmful_Cultural_Practices_in_the_West__Women_and_Psychology_1.pdf, p. 24. Accessed 27 March 2022.

Jeffreys, Sheila. 2020. Men's sexual rights versus women's sex-based rights. *WHRC Webinar*, 18 April 2020 [Online]. https://www.womensdeclaration.com/documents/59/Mens_sexual_rights_versus_womens_sex.pdf. Accessed 29 March 2022.

Kinsey, A.C., Pomeroy, B.W., and Martin, C.E. 1948. *Sexual Behaviour in the Human Male* [Online]. Bloomington and Indianapolis: Indiana University Press. https://www.google.co.uk/books/edition/Sexual_Behavior_in_the_Human_Male/_3SHCgAAQBAJ?hl=en&gbpv=1&dq=most+males+restrict+themselves+a+limited+series+of+techniques+to+which+they+are–been+erotically+conditioned&pg=PA510&printsec=frontcover. Accessed 27 March 2022.

Lerner, Gerder. 1986. *The Creation of the Patriarchy*. Oxford: Oxford University Press.

Llewellyn-Jones, Lloyd (ed.). 2002. *Women's Dress in the Classical Greek World* [Online]. London: Duckworth and Classical Press of Wales. https://www.google.co.uk/books/edition/Women_s_Dress_in_the_Ancient_Greek_World/2BctEAAAQBAJ?hl=en&gbpv=1&dq=Hera+of+the+white+arms&pg=PA1&printsec=frontcover. Accessed 27 March 2022.

Lockley, Mike. 2019. Businesswoman calls for £5k sex dolls to be made available for free on the NHS. *Birmingham Mail*, 16 November 2019 [Online]. https://www.birminghammail.co.uk/news/midlands-news/businesswoman-sex-dolls-available-nhs-17256873. Accessed 27 March 2022.

Love Doll *Sex Doll Anus Repair, TPE Filler*. https://lovedoll.co.uk/video/sex-doll-anus-repair-tpe-filler/. Accessed 27 March 2022.

MacKinnon, Catherine. 1993. *Only Words* [Online]. Harvard University Press. https://foundationsofgenderstudies.files.wordpress.com/2013/01/catharine-mackinnon-only-words.pdf. Accessed 27 March 2022.

McDonald review of *OVID'S HOMER: Authority, Repetition, and Reception* (2018) by Weiden-Boyd, Barbara. [Online]. https://classicsforall.org.uk/reading-room/book-reviews/ovids-homer-authority-repetition-and-reception. Accessed 27 March 2022.

McInerney, I. 2022. Bodybuilder obsessed with sex dolls says 'it's like necrophilia' but cops can't stop him. *Daily Star*, 7 January 2022 [Online]. https://www.dailystar.co.uk/news/weird-news/bodybuilder-obsessed-sex-dolls-says25881268?fbclid=IwAR0pcxaMUr5S5jqFGwz4RTRnQ1crimfeGRtkY184a--0XPtQSU8ZMLIaxaw. Accessed 27 March 2022.

McKinnon, Catherine. 1993. *Only Words*. Massachusetts: Harvard University Press [Online]. https://foundationsofgenderstudies.files.wordpress.com/2013/01/catharine-mackinnon-only-words.pdf. Accessed 28 March 2022.

Meet the proud owner of a DOZEN sex dolls Mar 9. 2019. YouTube video, added by The Sun [Online]. https://www.youtube.com/watch?v=10MhpDAAtjo. Accessed 27 March 2022.

Michael, T. 2016. Sex robots modelled on your favourite celebs to take over the market, expert warns. *News.com.au*, 5 December 2016 [Online]. https://www.news.com.au/technology/gadgets/sex-robots-modelled-on-your-favourite-celebs-set-to-take-over-the-market-expert-warns/news-story/c6d3b0754933a20f22851e663b822f3b. Accessed 27 March 2022.

Mulvey, Laura. 1975. Visual and Other Pleasures. *Screen, 16*(3), Autumn 1975, Pages 6–18 Available at: https://doi.org/10.1093/screen/16.3.6. Accessed 12 November 2022.

Mulvey, Laura. 1989. *Visual and Other Pleasures*. Basingstoke: Palgrave.

Naughton, J. 2020. It's a sign of a broken system when only credit card firms can force Pornhub to change. *The Guardian*, 19 December 2020 [Online]. https://www.theguardian.com/commentisfree/2020/dec/19/pornhub-abuse-videos-new-york-times-mastercard-visa. Accessed 27 March 2022.

Nuffield Council on Bioethics 'Cosmetic Procedures: Ethical Issues. 22 June 2017 [Online]. Cosmetic procedures—The Nuffield Council on Bioethics (nuffieldbioethics.org). Accessed 27 March 2022.

Ovid. 2004a. *Metamorphoses, Orpheus' Song: Pygmalion* (trans. D. Raeburn). London: Penguin.

Ovid. 2004b. *Metamorphoses, Orpheus' Song: The Cerastae and Propoetides* (trans. D. Raeburn). London: Penguin.

Paludi, M.A., and Denmark, F.L. 2010. *Victims of Sexual Assault and Abuse: Resources and Responses for Individuals and Families. Volume 1 Incidence and Psychological Dimensions* [Online]. Oxford: Praeger. https://www.google.co.uk/books/edition/Victims_of_Sexual_Assault_and_Abuse_Reso/cuMSpZ2UdjgC?hl=en&gbpv=1&dq=sex+renders+violence+invisible&pg=PA340&printsec=frontcocer. Accessed 27 March 2022.

Parsons, J. 2020. Pornhub deletes 80% of videos after claims site is 'infested' with child abuse. *Metro*, 16 December 2020 [Online]. https://metro.co.uk/2020/12/16/pornhub-deletes-80-of-videos-after-claims-site-is-infested-with-child-abuse-13759151/. Accessed 27 March 2022.

Pornhub blog. 2021. https://www.pornhub.com/blog/11422. Accessed 27 March 2022.

Roberts, Sophie. 2019. Sex Doll customers are ordering models that look like their mate's girlfriends. *Mirror*, 24 August 2019 [Online]. https://www.mirror.co.uk/news/uk-news/sex-doll-customers-are-ordering-18996096. Accessed 27 March 2019.

Ruiz, Karen. 2020. I saw it and started to cry uncontrollably': Mom's horror after her eight-year-old daughter's image is stolen and turned into a child sex doll sold on Amazon for $559 as Elizabeth Smart calls for the sick toys to be banned. *Daily Mail* [Online]. https://www.dailymail.co.uk/news/article-8698431/Florida-mom-finds-child-sex-doll-Amazon-resembling-photo-8-year-old-daughter.html. Accessed 27 March 2022.

Safe Schools Alliance. Porn and Young People: The facts. https://safeschoolsallianceuk.net/2021/04/11/porn-and-young-people-the-facts/. Accessed 27 March 2022.

Sampson, Rachel. 2015. Film Theory 101—Laura Mulvey: The Male Gaze Theory. *Film Inquiry*, 27 October 2015 [Online]. https://www.filminquiry.com/film-theory-basics-laura-mulvey-male-gaze-theory/. Accessed 27 March 2022.

Shakespeare, Tom. 2016. A Point of View: Does Comestic Surgery Really Make People Feel Better About Their Bodies. *BBC News*, 22 January 2016. https://www.bbc.co.uk/news/magazine-35380469#. Accessed 12 November 2022.

Sheila Jeffreys Keynote Lecture: Fetishism and the Construction of Male Sexuality (17 July 2020) YouTube video, added by the Campaign Against Sex Robots [Online]. https://www.youtube.com/watch?v=oZ_zfMwk_RQ. Accessed 27 March 2022.

Silicone sex—When men love dolls. 2019a. YouTube video, added by DW Documentary [Online]. https://youtu.be/VUAf_e9Y6XA. Accessed 27 March @18.55 mins.

Silicone sex—When men love dolls. 2019b. YouTube video, added by DW Documentary [Online]. https://youtu.be/VUAf_e9Y6XA. Accessed 27 March.

Vera-Gray, Fiona., McGlyn, Clare., Kureshi, Ibad, Butterby, Kate. 2021. Sexual violence as a sexual script in mainstream online pornography. *The British Journal of Criminology*, 61 (5): 1243–1260 [Online]. https://academic.oup.com/bjc/article/61/5/1243/6208896. Accessed 27 March 2022.

Woman Makes Custom Sex Dolls for Bereaved Men 18th June 2019 YouTube video, added by This Morning [Online]. https://www.youtube.com/watch?v=qubm_JiBfyo. Accessed 27 March 2022.

Wosk, Julie. 2015. *My Fair Ladies: Female robots, androids and other artificial Eves*. New Brunswick: Rutgers University Press.

Wright, P.J, Herbenick, D, and Tokunaga R.S. 2021. Pornography consumption and sexual choking: an evaluation of theoretical mechanisms. *Health Communication* [Online]. https://www.tandfonline.com/doi/abs/10.1080/10410236.2021.1991641?journalCode=hhth20. Accessed 27 March 2022.

CHAPTER 3

Mapping the Uses of 'Sex' Dolls: Pornographic Content, Doll Brothels and the Similarities with Rape

Lydia Delicado-Moratalla

"*The slave is a robot: its behaviour determined by the interest of another, its will by the will of another, its body functioning as a vehicle of another*".
(Marilyn Frye 1983: 65)

Introduction

The ways in which the sexual subordination of women to men is constructed within the structures of patriarchal societies has been a very important field of study in feminist theory since the 1960s (Dworkin

L. Delicado-Moratalla (✉)
Lecturer of Human Sciences and Education, Public University of Navarra, Navarra, Spain
e-mail: lydia.delicado@unavarra.es

© The Author(s), under exclusive license to Springer Nature Switzerland AG 2022

K. Richardson and C. Odlind (eds.), *Man-Made Women*, Social and Cultural Studies of Robots and AI,
https://doi.org/10.1007/978-3-031-19381-1_3

1974; Jeffreys 2012 [1991]; Millett 2017[1969]; Cobo 2017). An important part of feminist inquiries has been dedicated to the sexual objectification of women within the pornographic industry (Frye 1983; Jeffreys 2009; Walter 2010; Cobo 2011, 2020).

In the last decade, there has been a rise in the production and commercialisation of 'sex' dolls[1] (Kleeman 2017; Olucha 2019). These objects have become increasingly popular, and a variety of virtual communities have been appearing on the internet (Midleweek 2020; Delicado-Moratalla 2021).

While advocates of pornographic dolls and robots (Levy 2017; Pearson 2015; Danaher and McArthur 2018) claim that societies must get used to the idea of building love relationships with objects and have thus created a rationale to justify the normalisation of hyper realistic female replicas in everyday life, a simple search on the internet with the words 'sex doll' offers millions of images that are distanced from loving or caring behaviours (Delicado-Moratalla 2021). Although the creators of (robotic) 'sex' dolls (see declarations of RealDoll creator in Kleeman 2017; or promotional content of www.dsdollrobotics.com) argue that their aim is to create opportunities for companionship, any *Google* search will show a myriad of pornographic, often sadistic, content based on the use of real dolls to simulate humiliating treatment against women (Delicado-Moratalla 2021).

Given that, in this chapter I will map specific male uses of 'sex' dolls in order to build empirical knowledge about some of the effects caused by the introduction of these objects in everyday life. To do so, I will navigate through several virtual communities of real doll users and analyse the case of some 'sex' doll brothels. I will visit specific porn genres dedicated to 'sex' dolls and depict observed behaviours that particularly link with rape scenarios. Finally, I will suggest a feminist understanding of all the examined material.

[1] Although the industry and consumers refer to these devices as sex robots, sex dolls, sexbots or robotic sex dolls, I will follow Kathleen Richardson's term (this book) and use the expressions pornographic robots, porn robots, porn dolls and robotic porn dolls. To acknowledge my disagreement with the industry's designations, I will refer to these objects as 'sex' dolls/robots or 'sexbots.' I will also treat porn dolls and porn robots indistinctly, because dolls are the antecedents of robotic dolls. Both have very similar uses and have been designed for the same purposes.

The 'Sex' Doll/Robot Rationale

Definitions of 'sex' dolls and robots offer a very clear message for readers: "sex robots are robots that are used for having sex" (Bendel 2021: 1) or a 'sex' robot "is any artificial entity that is used for sexual purposes" (Danaher 2018: 4). According to Bendel (2021: 2), an inflatable doll would be a "sex partner" for men who cannot afford a more sophisticated object or prefer an old school style doll. Furthermore, Levy (2017: 3) refers to 'sex' robots as "partner robots."

Doll/robot advocates often claim that 'sex' dolls and robots are suitable for companionship or even for having an intimate relationship. David Levy (2008), the author of *Love and Sex with Robots,* has also written in defense of marrying robots, which he considers could be "sufficiently human-like and sufficiently appealing in various ways to take on the role of a partner in a relationship with a human being" (Levy 2017: 3).

Matt McMullen, CEO of *Abyss Creations* and founder of *RealBotix*, believes that a 'sex' robot "is much more about companionship" (Olucha, 2019) and that the robot Harmony would know how many times its owner likes "to have sex [with it]" (Kleeman 2017). At the same time, *Ai-Tech* advertises their product with words like "emotional care" and "intelligent robot girlfriend" (www.ai-aitech.com). The futurologist Ian Pearson further speculates on robots making "love to people physically" and giving "the best sex experience" they have ever had (Olucha 2019).

Particularly noteworthy are the narratives suggesting that 'love' dolls/robots are becoming replacements of women and have been referred to as "artificial love servants" (Bendel 2021: 6). Hauskeller (2018: 203–204) opts for the expression "automatic sweethearts" and wonders whether a "sexual companion robot" could be more satisfactory than a human lover. He elaborates on the idea of love and sex with "robotic persons" (Hauskeller 2018: 208) and their 'ability' to love. Additionally, Nyholm and Frank (2018: 219) explore the likelihood of "mutual love with a robot." Although they recognise that mutuality cannot happen with dolls, they support the idea of "mutual romantic love between humans and advanced sex robots" (Nyholm and Frank 2018: 219).

Interestingly, none of these ideas make reference to masturbation. All the approaches described above share a common element: they provide a rationale for legitimising the acceptance of objects that mimic—mainly—female humans and acknowledge the (robotic) dolls as companions for intimacy, love and sexual relationships in everyday life. Their ideas have built up discourse that has facilitated the development of 'sex' robots.

The Rise of Doll Pornography

According to Langcaster-James and Bentley (2018), consumers of 'sex' dolls particularly enjoy designing scenes, dressing dolls, taking pictures and distributing their pornographic material among virtual communities. On the blog www.amordeplastico.com the content creator writes about how he designs every detail of what he does with his dolls and how he presents them. He buys clothes, lingerie and accessories for his objects. A similar kind of pleasure can be observed in the ways a particular doll consumer, who appears in a documentary about the release of the 'sex' robot Harmony in California (Sweeney 2018), styles his dolls. Doll users design elaborate settings and put the objects in poses to photograph or video and then upload them onto social media to share.

There has been an explosion of virtual communities in just the last few years. The *RealDoll* online forum counts more than 11.4 million followers. *The Doll Forum* has over 50,000 followers (Middleweek 2020), the *Sex Doll Channel 2* on *Youtube,* which hosts more than 300 videos, recorded 16,000 subscribers in February 2021 and 32,400 in March 2022. The blog www.amordeplastico.com has 1,980 followers on Twitter. Berlin's 'sex' robot brothel *Kokeshi* by *Cybrothel* (@kokeshiberlin) has an *Instagram* account with over 2,000 followers (data from March 2022).

Specific porn genres have been created following the rise of 'real' dolls and 'sex' robots in the last decade. There is a wide range of media coverage and a variety of porn products have derived from what men do to their dolls. For example, Coverdoll (www.coverdoll.com) is a digital magazine that publishes a monthly issue and www.amordeplastico.com has announced the release of a book full of doll porn images that celebrates the anniversary of a doll company.

Links dedicated to the genre of 'sex' robot porn are part of the videos distributed on mainstream online platforms: *Xvideos* houses around 6,500 'sex' doll videos and over 200 videos related to 'sex' robots; the most viewed "sex doll" video has had 35 million visits to date (Delicado-Moratalla 2021). The channel www.pacoanddolls.com in *Pornhub* has 8,000 subscribers and has had 11,5 million visits to its videos (data from November 2021). *Pornhub* also hosts channels specifically dedicated to rape. When carrying out research on the *Pornhub* platform in February 2021, the title "*violación*" (rape) hosted 10 videos that totalled 43,800 visits (Delicado-Moratalla 2021). The words tagged "*muñecas reales*" (real dolls) contained 1367 videos with women acting as dolls. Disturbingly, these images present sexual violence against women (Delicado-Moratalla 2021). Producers of 'sex' doll/robot' porn are clearly looking to simulate rape and battery, which is also a trending topic in mainstream online porn (Whisnant 2016; Alario Gavilán 2018).

Considering all the information above and that "every second 28,258 internet users are viewing porn; 25% of all search engine requests are pornography related; and 35% of all internet downloads are pornographic," (Hall 2017: 130), the potential influence these trends can have on sexuality is not negligible. It is in fact a very important issue to take into consideration.

Brothels with Dolls and Dungeons

In February 2017, a 'sex' doll brothel was opened in Barcelona, the first of its kind to be located in Europe. Later, *Lumidolls* expanded their business and opened brothels in Turin, Moscow and Nagoya (www.lumidolls.com). They advertise: "the first real dolls online shop with brothels where you can try out the dolls before making them your companions […] live your fantasies without any limits in our facilities." As an aside, 'lumi' is a code word used by prostitution buyers in Spain to refer to prostituted women on blogs like www.pasion.com, where they rank, recommend or vilify women.

Lumidolls markets every 'sex' doll with a 'personal' profile. Each doll offers a pornographic narrative together with a set of images to trigger consumers' sexual fantasies. Information on the website is organised so that 'sex' dolls can be selected according to an artificial life story that is based on nationality, ethnicity, skin colour, age and body features.

Significantly, attention is directed to the body parts that are commonly exploited in pornography and prostitution: bottom, breasts, genitals and mouth. Under the label "teens," the doll *Chi* represents a female that looks like a teenager. The text says: "men can be gross sometimes, but women are always tender and delicate."

In the context of the commercialisation of sex dolls, some marketeers are trying to normalise their use within couples. For example, the description accompanying *Chi* declares "while your wife kisses me on my mouth you could eat my ass." The message is apparently effective. In the

Sex Robot Documentary (Sweeney 2018), the protagonist, who has three dolls, expresses his determination to bring a 'sex' robot to the home he shares with his wife.

A number of television programs have explored 'sex' doll brothels. The BBC (2019) visited Turin's *Lumidolls*, where a prostitution buyer gave a positive review: "A prostitute is a real person, she can judge you for your fantasies. A doll cannot do the same. With a doll, everything I have to think about is my own satisfaction. I felt free." These ideas are also shared by the brothel manager: "I sell magic. The client who comes here wants to disconnect from everything. He does not feel like making any effort with a girl. He does not even want to talk to a girl. He comes here and releases his stress." In these testimonies, there is no empathetic sexuality. Sexual gratification is self-centred and there is an explicit narrative that avoids intimate contact, dialogue or reciprocity with a woman. These characteristics resonate with normative male heterosexual behaviour, as I will evidence below.

Similar elements are found in the video reports available in the media regarding doll brothels. In *Lumidolls* Barcelona, which offers prostituted women and real dolls to consume at the same location, a male reporter from the BBC (Eastman 2018) is introduced to *Brandy*: "the doll has got three holes, like women, so you know how to use it" its pimp tells the reporter. Women are understood as passive bodies with three holes to be penetrated by active men. This imagined sexual setting implies that sexuality and sexual intercourse are believed to be unidirectional, not mutual, a scene where men dominate and penetrate women using what is understood to be holes and not human organs or parts of the female body. Mirroring real female body parts with holes indicates a dehumanised idea of women, who are, at the same time, represented as submissive empty objects that men use for their self-focused sexual pleasure.

In *Bordoll*, a dolls' brothel located in Dortmund, a dominatrix pimps out dolls and advertises, as a main feature of the facility, the availability of a complete range of instruments of torture: "dolls are not sentient beings, and they look good, they have no illness, they attend all kind of services without complaining and consumers can do whatever they like with them […]. Men can be selfish, they do not need to worry about the dolls' feelings" (Winter 2018). Her discourse normalises, privileges, reproduces and reinforces male sexual selfishness and domination. As Eva Illouz (2014) explains, male sexuality needs to reproduce patterns of commitment-phobic, anti-emotional sex in order to avoid proximity with femininity. To show and to free emotions or feelings in sexuality is part of the gendered construction of femininity and it is seen to be equivalent to weakness. So, men pay attention to the opposite values of femininity when building and developing their sexual behaviour, in order to maintain their positions of power. This explains what dolls' brothel consumers search for and speak about.

LAD Madrid has been inspired by other doll brothels and gives customers the opportunity to 'try' dolls before buying them The distinctiveness of this brothel is that its facilities comprise two dungeons fully equipped for performing sadistic 'sexual' acts. It is interesting to note that sadomasochism is now mainstream in porn (Seida and Shor 2019; Fritz et al. 2020). The owner of *LAD* Madrid explains his business purpose: "dolls do not complain… Clients also come here looking for companionship. They do not dare to have intimacy with a person, they know they do not need to chat or interact with dolls and that makes them feel comfortable" (The Objective 2019). These actions produce and reinforce dehumanisation in male consumers' sexuality.

The sadistic use of dolls is not only embraced in these brothels. Apparently, people surveyed have similar thoughts when imagining a future with 'sexbots.' In one study by Tufts University (Scheutz 2016 cited in Wagner 2018: 55) "people rank the idea of using a sexbot to engage in 'rough sex or sadistic behavior' as a 5.23 out of 7 for acceptable behavior … (On a scale where a 1 is unacceptable and a 7 is completely acceptable)."

Throughout all the testimonies above, it is clear that avoiding intimate interaction, mutuality and communication with women is central to the production and use of 'sex' dolls /robots. Although what is in play are plastic replicas of female bodies, the imagery always pivots around women and girls. The misogynistic narratives produced by the managers of the doll brothels are explicit and reformulate predominant patriarchal heterosexual patterns. Furthermore, it can be seen that both the consumers' and the pimps' ideas of heteronormative male sexuality are built not only upon a recreational notion of sex (Illouz 2014) but also upon a worldview of domination and subordination, together with the sexual objectification of women, which are characteristic of patriarchal violence.

Torture against an imitation of a woman appears absolutely normalised; it is part of a panoply of leisure activities for having fun. The representation of women's unlimited subordination to men's sexual fantasies and the simulation of women's acceptance of torture and pain as sexually appealing appear to be an important business target in these doll brothels. The uncritical approach to torture performed within the sexual realm reduces its violence and depoliticises the practice.

The Simulation of Doll Rape

As Millett (2017[1969]) and Dworkin (1974 2015[1981]) have argued, rape has been romanticised by pornography since the sexual revolution. Rape became a core narrative in pornography and was reinforced by the popularisation of sadomasochism in the second half of the twentieth century (Jeffreys 2012[1991]). Particularly relevant is the fact that "violent pornography represents women deriving sexual pleasure from rape" (Eaton 2017: 683). Brutal rapes have been normalised in pornography and gang-bang rapes are easy to find on any online platform of 'adult' content (Jensen and Okrina 2004; Boyle 2010; Alario Gavilán 2018).

Frequently reported in the news are cases of women and girls being raped while unconscious. For example, Ayuso (2021) presented the case of a woman in her sixties who had been repeatedly raped while forcibly unconscious. Her husband had been drugging her in order to prostitute her. He contacted men online and pimped his wife. Pictures and images were taken while her multiple rapes were taking place and, afterwards, they were uploaded and distributed online. Cases of teenage girls found unconscious and raped in parks (Cabanes 2021) and news about gang-bang rapes which were perpetrated after having drugged girls (González 2021) are not uncommon. The Ministry of Justice has reported that drugs were used on one in three of the victims of sexual assault in Spain in the last five years (EuropaPress 2022).

The calculated unconsciousness commonly found in rape settings is the element that leads me to make tentative connections with the scenes where men use 'sex' dolls, because dolls mimic dead or unconscious female bodies.

When conducting research on blogs like amordeplastico.com, the channel pacoanddolls.com in *Pornhub*, coverdoll.com magazine and *The Sex Doll Channel 2* on *YouTube*, observation of the content shows that consumers of dolls have constructed a sexual narrative in which the presupposed female actor—the doll—is not expected to be a participant. The female replica is expected to act as an immobilised frozen body while being penetrated. Sometimes dolls even appear with their arms chained. The scenes not only mimic the sexual subordination of women to men, but also show that all control is held by the male actor with no possibility for the female to object. Doll consumers act as if 'sexually' using completely realistic representations of unconscious women—and sometimes underage girls—is acceptable.

With the pornography that doll consumers are distributing, images that simulate rape become public—they are effectively elaborating a narrative of acceptance of rape. This metaphoric use of rape representations is problematic because it sets the conditions for accepting rape as something desirable. These men are sharing and uploading sets of pro-rape images, fantasies that travel through internet networks without any critical approach. These rape-devotional sexual masculinities become thus visible, popular and mainstream, creating room for their validation and constructing unsafe social spaces for women.

On the internet websites I have referred to in this section, it is common to find comments and conversations about the consumers 'right' to play with their dolls. They defend the idea that nothing harmful is taking place, because they are not hurting a real girl or woman. Perhaps they have not thought about the links between rape and what they are performing. Although it might not be immediately obvious to them, they are, in fact, harming women indirectly because they are creating a narrative that builds from the symbolic abuse of women. And this content has the ability to influence others, because it is being reproduced hundreds of times with every visualisation and download.

Final Thoughts and Ideas

Patriarchal logic has produced devices and artefacts that encourage the perpetuation of structures of male dominance over women. 'Sex' dolls and robots are an extreme expression of the objectification of women that legitimates a rhetoric of male abuse of power in sexuality.

Virtual activities related to porn dolls shape a specific geography within cyberspace. The websites researched in this paper depict images and films of men practising sexual domination over hyper realistic female replicas.

An analysis of pornographic websites, platforms and social media confirms that consumers of 'sex' dolls are enthusiastically producing and distributing porn. Data on the issue indicate it is a growing market, gaining acceptability and followers through time. Dolls and robots get similar treatment to humans regarding their profiles on social media, amassing thousands of followers in the cases discussed here.

'Sex' dolls and porn robots encourage the mechanisation of sexuality and narrow the sexual universe to a pornographic one of repeated acts of masturbatory penetration into plastic replicas of women and girls. They orientate sexuality away from mutuality towards non-empathetic, non-emotional, non-human acts, reproducing and reinforcing gender roles and stereotypical behaviours. This is problematic because it assumes that sexuality can be even further distanced from human reciprocal acts, and cannot be aligned with feminist sexual politics.

The practices that are taking place with 'sex' robots and dolls, like the development of brothels, pornography and virtual communities, are normalising torture against women as pleasant and sexually arousing. In my opinion, the acceptability of these practices as "only on dolls" trivialises violence against women and girls. Both the idea and the practice of torture against women or the imitation of women, collide with the feminist project of overcoming patriarchal injustice.

A feminist sexuality aims to eroticise equality (Jeffreys 1993) and understands mutuality and reciprocity as being at the very core of an intimate relationship. Domination and subordination belong to hierarchies, so they cannot be part of the feminist construction of sexuality. Inflicting pain, practising acts of violence and encouraging sadomasochist behaviours are part of the structure of domination. Building a feminist sexuality is founded in emotional consideration, egalitarianism and shared mutual pleasure.

References

Ayuso, Silvia. 2021. Una sexagenaria francesa fue drogada por su marido y violada durante una década por decenas de hombres. *El País*.

BBC News. 2019. Sex Doll Brothels: A Growing Trend? *BBC News*

Bendel, Oliver. 2021. Love Dolls and Sex Robots in Unproven and Unexplored Fields of Application. *Paladyn* 12 (1): 1–12. https://doi.org/10.1515/pjbr-2021-0004.

Boyle, Karen. 2010. *Everyday Pornography*. Oxon: Routledge.

Cabanes, Ignacio. 2021, November 14. Detenido por violar a una menor de 14 años tras quedar por Instagram. *Levante*.

Cobo, Rosa. 2011. *Hacia una nueva política sexual. Las mujeres ante la reacción patriarcal*. Madrid: Catarata.

Cobo, Rosa. 2017. *La prostitución en el corazón del capitalismo*. Madrid: La Catarata.

Cobo, Rosa. 2020. *Pornografía. El placer del poder*. Barcelona: Penguin Random House Grupo Editorial.

Danaher, John. 2018. Should We Be Thinking about Robot Sex? In *Robot Sex. Social and Ethical Implications,* ed. J. Danaher and N. McArthur, 3–14. Cambridge: The MIT Press.

Danaher, John, and Neil McArthur, eds. 2018. *Robot Sex. Social and Ethical Implications*. Cambridge, MA: MIT Press.

Delicado-Moratalla, Lydia. 2021. Los discursos y las prácticas de cosificación de las mujeres en la era de internet: las muñecas sexuales en el ciberespacio. In *Feminismo digital. Los derechos de las mujeres y brecha sexista en internet,* ed. Tasia Aránguez and Ozana Olariu, 486–503. Dykinson.

Dworkin, Andrea. 1974. *Woman Hating*. New York: Penguin Books.

Dworkin, A. (2015[1981]). *Pornography: Men Possessing Women*. New York: Plume. Penguin Group.

Eastman, John. (2018). *The Future of Sex? Sex Robots and Us*. BBC.

Eaton, A. W. 2017. A Sensible Antiporn Feminism. *Aesthetics: A Reader in Philosophy of the Arts: Fourth Edition*, 117: 470–474. https://doi.org/10.4324/9781315303673.

EuropaPress. 2022, March 8. *Justicia estima que una de cada tres agresiones sexuales desde 2017 se producen bajo el efecto de drogas o alcohol*. Europa Press.

Fritz, N., Malic, V., Paul, B., & Zhou, Y. 2020. A Descriptive Analysis of the Types, Targets, and Relative Frequency of Aggression in Mainstream Pornography. Archives of Sexual Behavior, 49 (8), 3041–3053. https://doi.org/10.1007/s10508-020-01773-0

Frye, Marilyn. 1983. *The Politics of Reality: Essays in Feminist Theory*. New York: Crossing Press.

Gavilán, Alario, and Mónica. 2018. La influencia del imaginario de la pornografía hegemónica en la construcción del deseo sexual masculino prostituyente: Un análisis de la demanda de prostitución. *Asparkía. Investigació Feminista* 33: 61–79. https://doi.org/10.6035/Asparkia.2018.33.4.

González, Germán. 2021. *Los Mossos apuntan a una violación en grupo de la menor de Igualada*. El Mundo.

Hall, Lynne. 2017. Sex With Robots for Love Free Encounters. Lecture Notes in Computer Science (Including Subseries Lecture Notes in Artificial Intelligence and Lecture Notes in Bioinformatics). *LNAI* 1023: 7128–7136. https://doi.org/10.1007/978-3-319-57738-8_12.

Hauskeller, Michael. 2018. Automatic Sweethearts for Transhumanists. In *Robot Sex. Social and Ethical Implications,* ed. J. Danaher and N. McArthur, 203–218. Cambridge: The MIT Press.

Illouz, Eva. 2014. *Erotismo de autoayuda: 50 sombras de Grey y el nuevo orden romántico*. Madrid: Katz Editores.

Jeffreys, Sheila. 1993. *The Lesbian Heresy: A Feminist Perspective on the Lesbian Sexual Revolution*. Chicago: Spinifex.

Jeffreys, Sheila. 2009. *The Industrial Vagina: The Political Economy of the Global Sex Trade*. Abingdon: Routledge.

Jeffreys, Sheila. 2012[1991]. *Anticlimax: A Feminist Perspective on the Sexual Revolution*. Spinifex Press.

Jensen, Robert and Debbie Okrina. 2004. Pornography and Sexual Violence. *Applied Research Forum*, July, 1–8.

Kleeman, J. 2017. *Rise of the Sex Robots*. United Kingdom: The Guardian.

Langcaster-James, Mitchell, and Gillian R. Bentley. 2018. Beyond the Sex Doll: Post-Human Companionship and the Rise of the "Allodoll." *Robotics* 7: 4. https://doi.org/10.3390/robotics7040062.

Lerner, Gerda. 1990. *La creación del patriarcado*. Barcelona: Crítica.

Levy, David. 2008. *Amor y sexo con robots*. Paidós Ibérica.

Levy, David. 2017. Why Not Marry a Robot? Lecture Notes in Computer Science (Including Subseries Lecture Notes in Artificial Intelligence and Lecture Notes in Bioinformatics). *LNAI* 10237: 3–13. https://doi.org/10.1007/978-3-319-57738-8_1.

Linden, Robyn Ruth et al. (eds.). 1982. *Against Sadomasochism: A Radical Feminist Analysis*. Frog in the Well.

Middleweek, Belinda. 2020. Male Homosocial Bonds and Perceptions of human–Robot Relationships in An Online Sex Doll Forum. *Sexualities*, 24(3), 370–387. https://doi.org/ https://doi.org/10.1177/1363460720932383

Millett, Kate. 2017[1969]. *Política sexual*. Madrid: Cátedra.

Nyholm, Sven, and Frank, Lily Eva. 2018. From Sex Robots to Love Robots: Is Mutual Love with a Robot Possible? In *Robot Sex. Social and Ethical Implications,* ed. J. Danaher and N. McArthur, 219–244. Cambridge, MA and London, England.: The MIT Press.

Olucha, R. 2019. Sexe digital i amor programat. TV3.

Pearson, Ian. 2015. *The Future of Sex Report. The Rise of the Robosexuals*. http://graphics.bondara.com/Future_sex_report.pdf

Seida, K., & Shor, E. 2019. Aggression and Pleasure in Opposite-Sex and Same-Sex Mainstream Online Pornography: A Comparative Content Analysis of Dyadic Scenes. *Journal of Sex Research*, 00 (00), 1–13. https://doi.org/10.1080/00224499.2019.1696275

Sweeney, Nick. 2018. *Sex Robot Documentary*. Raw TV Channel 4.

The Objective. 2019. *Dentro del primer "burdel" de muñecas de Madrid*. Spain: The Objective.

Wagner, Christian. 2018. Sexbots: The Ethical Ramifications of Social Robotics' Dark Side. *Ai Matters* 3 (4): 52–58. https://doi.org/10.1145/3175502.3175513.

Walter, Natasha. 2010. *Muñecas vivientes. El regreso del sexismo*. Madrid: Turner Publicaciones.

Whisnant, Rebecca. 2016. Pornography, Humiliation, and Consent. *Sexualization, Media, & Society* 2 (3): 237462381666287. https://doi.org/10.1177/2374623816662876.

Winter, Chase. 2018. *Inside Bordoll, a German sex-doll brothel in Dortmund*. DW.

CHAPTER 4

Fetishism and the Construction of Male Sexuality

Sheila Jeffreys

In this chapter, I explore why male sexuality is constructed in such a way that men can get sexually excited by plastic dolls, sex robots or chatbots, as well as a panoply of other fetish objects which replace human beings. The fact that men can be sexually excited by inanimate objects needs to be explained. There is no evidence that women have this problem and sex dolls made to represent male bodies are not marketed to women. I argue that this interest in men can only be understood by examining the construction of male sexuality out of the power relations of male domination.

In the literature of sexologists, the scientists of sex, or in popular literature about sex, male sexuality is assumed to be biologically constructed

S. Jeffreys (✉)
University of Melbourne, Melbourne, Australia
e-mail: sheila@unimelb.edu.au

© The Author(s), under exclusive license to Springer Nature Switzerland AG 2022

K. Richardson and C. Odlind (eds.), *Man-Made Women*, Social and Cultural Studies of Robots and AI,
https://doi.org/10.1007/978-3-031-19381-1_4

and the idea that it might be constructed by forces of power is not considered (Jeffreys 1997 2008a 2008b). Sometimes no distinction is even made in research using human subjects between men and women. One example is a study of the prevalence of different forms of sexual fetishism with subjects found through online fetish support groups (Scorolli et al. 2007). It must be assumed that their subjects are all men because sexual difference and women are not mentioned. There is some literature which considers why fetishism, for instance, is very unusual in women except in relation to sadomasochism. The explanations offered do not consider the idea that sexuality could be affected by the relations of power between men and women in which it is formed, or that such relations of power exist (Dawson et al. 2016; Ventriglio et al. 2019).

Radical feminist theorists take a social constructionist approach to sexuality, but also a distinctly political one. We understand sexuality to be constructed out of the power relationship between the sexes. It would be very strange if it was not. Radical feminist researchers argue that the oppression of women is mediated and enforced through sexual practice (Mackinnon 1989; Jeffreys 1990/2011; Dworkin 1987). In heterosexual sex members of the two sex classes of male domination, women and men, engage in an intimate activity that involves the sexual organs on which their elevation or degradation in the hierarchy of sex class is founded. It is hard to imagine what could construct the male-dominant and hierarchical forms in which heterosexual sex takes place apart from this power relation.

Male Sexuality and the Social Control of Women

Understanding male sexuality and working out how to change it is not a matter of abstract interest for women. The way in which male sexuality is constructed dramatically affects women's lives and serves both to create and to maintain male domination. The exercise of the male sex right, men's right of sexual access to the bodies of women and children (Pateman 1988), has led to the creation of considerable industries of pain in which women and children are warehoused for men's delight in prostitution and pornography (Jeffreys 2008a 2008b). The male sex right means that men feel entitled to demand women's sexual attention or touch or use their bodies in public space, in the form of sexual harassment and rape (Romito 2008). It means that states and cultures organise and donate the bodies of women and girls to men for their exclusive use through marriage in its innumerable forms, temporary, arranged, forced, 'love' marriages, polygamy (Jeffreys 2012). In these arrangements of ownership men often still have absolute right to use women at will even if this prerogative has been limited in most jurisdictions since the 1970s through the existence of little used laws against 'marital' rape.

The expression of male sexuality through the practice of sexual harassment, stranger rape and sex murder, sets up siege conditions for women and girls (Larkin 1997). They have to manage their routes out of home and the way they negotiate a school or workplace, in order to avoid sexual assault or exploitation (Vera-Gray and Kelly 2020). The best example of this state of siege is the spycam emergency in Korea (Human Rights Watch 2021). In Korea, men armed with excellent technology are using spycams and mobile phones to lay siege to women. Men use mobile phones to photograph up women's clothing as they walk along the street, and they are used through and over toilet walls which are pockmarked with holes to enable the men access. The film is streamed straight onto the internet, often for profit. Criminal gangs stake out toilets and 'love hotels' to film women in sexual activity and they make money from streaming this. So serious is the situation that specialist firms have been set up which are paid to sweep premises such as conference centres to get rid of the spycams until the perpetrators can reestablish them. The perpetrators explain that their excitement comes from stealing the photos of the woman without her knowledge. The effect is that women are continually observed and striving to prevent themselves from being observed, and using the toilets is dangerous. The spycam situation demonstrates the extent to which each advance in technology can be and is used to advance men's sexual violation of women and the destruction of our freedom. It suggests the serious need for feminists to analyse male sexuality and work to change it.

Social Construction of Sexuality

The belief that sexuality is shaped by biology is still dominant in the discipline of sexology. Sexologists have done a great deal of research over the last three decades, for instance, to show that homosexuality is innate (Hamer and Copeland 1994; Levay 1991; Blanchard 2001), though feminist researchers have disagreed with them profoundly (Rose 1996; Kaplan and Rogers 2003). Some do not just consider sexual orientation to be formed by biology but child sexual abuse also, euphemistically called 'paedophilia' (Seto 2012). There is some disagreement over the etiology of men's fetish behaviour though, with many following Freud in seeing this as acquired through early childhood influences. There is little mention or consideration of why men might manifest paraphilias and women do not, however. The only suggested explanation is what they see as a biological one, that men have a higher 'sex drive,' though it is hard to see how this could explain such practices as nappy fetishism or transvestism.

Feminists see male sexuality as socially constructed by forces of power, not just by whether they got excited by their nursemaid's shoes when they were six years old, as traditional psychoanalysts might have it. We argue that sexuality consists of chosen behaviour that men have responsibility for and can change (Jackson and Scott 2010; MacKinnon 1989). Saying that sexuality is socially constructed means that though the capacity for sexual pleasure is biological, the way in which people behave sexually and the persons they choose to be sexual with, are shaped by culture and experience (Gagnon and Simon 1974). Feminist theorists argue that the main power involved in shaping the expression of sexuality is the power relationship of male domination and women's subordination. Sexuality is not a neutral form of behaviour such as whether you use a fork to eat or not, but rather the mechanism through which the power relations between the sexes are established and symbolically and actually acted out. Women learn to eroticise their own subordination, and men learn to eroticise their dominance (Jeffreys 1990/2011).

Radical feminist theorists have argued that sexuality is both formed from the power relations of male domination and forms its very foundation. Andrea Dworkin, for instance, in her examination of the dynamics of penis in vagina sexual intercourse, identified men's determination to penetrate women as constituting a political battle. Intercourse, she said, was a vital practice in achieving men's power over women, "Rebellion" in intercourse, she said, "is the death of a system of gender hierarchy premised on a sexual victory over the vagina. The triumphant fuck is virtually synonymous with masculinity" (Dworkin 1987: 149). Catharine MacKinnon argued that sexuality is central to the oppression of women. She says that sexuality is, "a social construct of male power defined by men, forced upon women, and constitutive of the meaning of gender" (1989: 128). She posits that sexuality is the organising principle of male supremacy. She points out that what makes male domination different from all other forms of oppression is that it is sexualised, "Male dominance is sexual. Meaning: men in particular, if not men alone, sexualize hierarchy; gender is one" and as a result, "The male sexual role… centers on aggressive intrusion on those with less power" (ibid.: 127, 128).

It may be a problem for boys and men as they learn to be sexual, that they have to work out how to relate in a way which may make them vulnerable and involve intimacy with a member of a group of people, women, that they are acculturated to consider inferior and unclean. This is a problem never mentioned, but many fetishes are concerned with men's disgust at women. For men who have this problem, women's genitals are likely to be seen as disgusting, a reasonable conclusion considering that these body parts are precisely what marks women out as an inferior sex class. Andrea Dworkin wrote in her book *Intercourse* about the way that men's disgust at women's genitals shapes the construction of sex. She points out that Freud suffered from this problem, explaining that his concept of the castration complex, and of how fetishistic behaviour was constructed, was based upon it. Freud decided, she said, that "just seeing those genitals turns a man gay or makes him rub up against rubber for a lifetime" (Dworkin 1987: 182). He had a problem with menstruation too, writing to Jung "Oh yes, I forgot to say that menstrual blood must be counted as excrement" (ibid.: 1987: 183). Dworkin argues that sex cannot be made "healthy" without ending the subordination of women because that is what constructs the idea that sex and or women are disgusting. As she explains, "Current dogma is to teach by rote that sex is 'healthy' as it if existed outside social relations, as if it had no ties to anything mean or lowdown, to history, to power, to the dispossession of women from freedom. But for sex not to mean dirt, for sex not to be dirty—the status of women would have to change radically" (ibid.: 173). It is entirely understandable that when faced with feeling sexually excited by the female genitals that revolt them, some men would associate excitement with revulsion and develop the perversion of coprophilia, for instance.

Sexology

The main educational influence in the construction of sexuality in the last two centuries has been sexology (Jeffreys 1990/2011). Sexology and sex therapy, a practice which applies its ideas, adjudicate on what is acceptable sexually, promote the male sex right and enforce women's obeisance to men's demands (Tyler 2011, 2016). Before the late nineteenth century in the west the church was in charge of sex. Miscreants were brought before the church courts for engaging in a prohibited activity such as anal sex. The church said what people were to do and with whom (Ingram 1990). By the second half of the century the church had lost its power to control the people, particularly working-class people in the towns who were unable to say who Jesus was when worried missionaries interrogated them. The maintenance of patriarchal authority over the power relations of sexuality moved to another authority, and this was sexology. In the late nineteenth century medical scientists took control of sex. They classified what the correct practice was, penis in vagina sex, and in the first part of the twentieth century created the requirement that women be enthusiastic about being penetrated. The doctors decided that between 40 and 100 per cent of women were in fact frigid (Jeffreys 1985/2003), and the main task for the sexologists thereafter was making sure women both allowed male-dominant and female-submissive penetration to take place and that they responded appropriately. But in their prescribing what normal heterosexual sex should be, the sexologists gave themselves the task of delineating the incorrect forms of sexual behaviour.

The Perversions

The sexologists made lists of the practices which were not what they saw as normal, originally called perversions and now more usually called paraphilias, a term which is seen as less judgemental. They collected examples of these more unusual behaviours from their male patients and gave them names. Richard von Krafft-Ebing described what he considered to be the perversions in his book *Psychopathia Sexualis* in 1887 (1931, first published 1886). The perversions were described in Latin so that only doctors could read about them. The names that Krafft-Ebing and other sexologists such as Henry Havelock Ellis and Sigmund Freud invented for the perversions were generally mixtures of Latin and Greek. The term capillary kleptomania, which is used by Ellis for men's practice of stealing women's hair by cutting off a plait and taking it away to masturbate into, for instance, combines terms of Greek and Latin origin (Ellis 1926). There was a very wide range of perversions recorded in sexological work in the early decades of the twentieth century. The perversions included renifleurism, whose male practitioners would gain satisfaction from the smell of women's or men's urine. In one case described in the literature a man would travel on a store elevator to enable bending forward to smell a woman's backside (Valentine 1953).

In the case of some of these scientists of sex, their enthusiasm for mapping the more outlandish sexual practices may have stemmed from their unusual sexual predilections. Henry Havelock Ellis, for instance, sought to normalise fetishism by naming it 'erotic symbolism,' a term which did not stick. In this he included urolagnia, or love of urine, which he practised himself in the form of getting women to urinate so that he could observe or listen (Jeffreys 1985/2003). Ellis took a very positive approach to 'erotic symbolism,' which was the practice of taking a part of the body or object associated with the body as an erotic object rather than another person, "By 'erotic symbolism' I mean that tendency whereby the lover's attention is diverted from the central focus of sexual attention to some object or process which is on the periphery of that focus" (Ellis 1926: 1). The practice, he asserted with confidence, signified intellectual sophistication, "in its less crude manifestations erotic symbolism easily lends itself to every degree of human refinement and intelligence" (ibid.: 3). Ellis reserved his greatest praise for his own favoured practice, urolagnia, stressing that it was "not uncommon" and "has been noted of men of high intellectual distinction," such, presumably, as himself (ibid.: 59). He does mention that Krafft-Ebing said that no women were fetishists and since women did not manifest these behaviours it might suggest, according to his prejudices, that women were less refined and intelligent. All that he included under the mantle of 'erotic symbolism' would now be called fetishism.

He considered erotic symbolism to be perfectly normal, being merely an exaggerated form of everyday sexual impulses, "It is important to remember, however, that while it becomes fantastic and abnormal in its extreme manifestations, it is in its essence absolutely normal" (ibid.: 8). In general, he thought, these behaviours were acquired rather than innate and the result of early experience: "the fetishisms are those which are most vaguely conditioned by inborn states of the organism and most definitely aroused by seemingly accidental associations or shocks in early life" (Ellis 1926, first published 1906: 28). But there were exceptions. In the case of foot and shoe 'erotic symbolism,' he saw the practice as innate. Using the example of the crippling of women in China through footbinding for men's sexual excitement, he argued that eroticising women's feet was an "occasional reappearance of a bygone instinct" which could be "said to arise on a congenital basis" as the "rare development of an inborn germ" (ibid.: 27).

Freud had a different explanation. He stated that a sexual fetish is a 'substitute for the penis' (Freud 1977: 353). In his account, boys see their mothers naked and realise they do not have penises. This causes a crisis of fear which makes them afraid that they will lose their own penis and this somehow makes them hate women's genitals. In self-defence they must create a fetish which serves as a substitute for the penis that their mother is seen to have mislaid. All men are affected by the horror of women's genitals, he said, "Probably no male human being is spared the fright of castration at the sight of a female genital" (ibid: 354). All men, then, are liable to develop fetishes to overcome their fear and shock.

Sex Dolls, Sex Robots, Chatbots

The sexologists did not name the form of male behaviour that involves using whole body substitutes for women such as in sex dolls or sex robots, or the practice of abusing women in a disembodied form through chatbots (Bardhan 2022). The technologies which would enable these practices, such as the invention of plastics and artificial intelligence did not exist when sexologists were most active in the creation of their indexes of perversions, but it is possible that they would have been included in the concept of the perversions and fetishism if they had. A version of a sex doll seems first to have been created in the mid-twentieth century as soon as the development of plastics made this possible, and it was called the 'sailor's friend.' It has not been given a specific name though there are discussions of some forms of behaviour which bear some resemblance. Some sexologists, however, have considered whether men's use of sex dolls might be classed with agalmatophilia, the sexual use of statues by men, or seen as a form of fetishism.

> Agalmatophilia is the pathological condition in which some people establish exclusive sexual relationships with statues. The condition is neither to be confused with pygmalionism nor with fetishism, although confusion sometimes arises about these three different manifestations of immature sexuality. (Scobie and Taylor 1975: 49)

Scobie and Taylor, who write about agalmatophilia, consider whether sexual interest in statues is similar to Pygmalionism, but think it is not because the agalmatophilist does not want to animate statues, "The myth of Pygmalion can apply only to those who actually bring statues to life, and not to those who use statues for their own sexual purposes without bringing them to life .... Pygmalion shunned marriage with real women because of their supposed vicious- ness; he fashioned an ideal woman for himself out of ivory and successfully appealed to Venus to animate the statue" (ibid). It does not resemble fetishism they consider because it is love of whole body statues rather than objects or body parts. For those feminists interested in the construction of male sexuality as a whole, of course, these differences are moot. The important question is why men divert their sexual and love interests onto anything that is not an adult human being of either sex, and prefer something inanimate or fashioned to their precise requirements instead.

The Liberation of the Perversions

Until the 1960s the sexologists still usually wrote about sexual perversions or paraphilias as mental health problems that required treatment. This changed in the late twentieth century, when many of the perversions were liberated and normalised. The so-called sexual revolution of the 1960s and 70s ushered in a wave of sexual liberalism which mainly consisted of ensuring men greater sexual access to girls and women and getting women to service their needs, including their paraphilias, without complaint. Alex Comfort's *Joy of Sex* from 1973 was an influential, sexually revolutionary, sex advice book, which advised women, who were seen as recalcitrant, to do all the things men liked (1979). His book is a how-to guide for women in the servicing of male sexual desires for sadomasochist sex, for fetishism and for other degrading practices, many of which they might want to reject as he acknowledges, "Most wives who don't like Chinese food will eat it occasionally for the pleasure of seeing a Sinophile husband enjoy it and vice versa" (1979: 11). There is no 'vice versa' in the book. Women could be annoyingly resistant to sadomasochism for instance, "Gagging and being gagged turns most men on – women profess to hate it in prospect" (Comfort 1979: 126). Comfort explains that "men wish that women's sexuality was like theirs, which it isn't. Male sexual response is far brisker and more automatic: it is triggered easily by things, like putting a quarter in a vending machine. Consequently, at a certain level and for all men, girls and parts of girls are, at this stimulus level, unpeople" (ibid: 34). He is confident that all men objectify women and treat them as not fully human and that fetishism is completely normal. The use of sex dolls was not mentioned but fits very well into his idea of natural male sexuality.

The sexual revolution created the basis, in its promotion of sexual liberalism and decensorship, for the development of the pornography industry (Jeffreys 1990/2011). As the new pornography industry developed it created niches to service all of the forms of fetishism that the sexologists had documented. Out of the sexual revolution there developed movements of sexual liberation, starting with the gay liberation movement but including, in the 1970s, the practice of child sexual abuse and then many of the perversions or paraphilias that the sexologists had taken such care to document decades before. The liberation of the perversions is called for by a book which was controversial but well known when it was published in English in 1966, Lars Ullerstam's *The Erotic Minorities: A sexual bill of rights*. It was originally published in Sweden, when Sweden and other Scandinavian countries such as Denmark, were seen in the counterculture of the 1960s as being in the forefront of the sexual revolution. Ullerstam was angry about what he called "old moralism" which had laws against men's expression of their perversions and fetishes, "to prevent exhibitionists, pedophiliacs, and certain kinds of scopophiliacs from ever being able to satisfy their sexual urges" (Ullerstam 1966: 11). He explains that what were once called 'perversions' must be destigmatised, because social disapproval of these forms of male sexual behaviour made men unhappy, "Throughout the ages the term 'perversion' has been applied to the most divergent phenomena. But generally, this poison label has been applied to human *needs*. When an individual is called a 'pervert' this means, as a rule, that he has the capacity for pleasure in a specific context. Whoever has as his concern the happiness of his fellow men ought therefore to appreciate and encourage 'perversions'" (Ullerstam 1966: 17). The 'perversions' he mentions are specifically male and mainly consisted of forms of sexual violence that harmed women and children. Scopophilia, for instance, consisted of the 'perversion' of getting sexual pleasure from sexual looking, and includes such practices as voyeurism and pornography. Pornography became legally and socially acceptable in succeeding years and was entirely normalised. In the case of some of his preferred perversions this was not so straightforward.

The model that the liberation movements used was that of the successful gay liberation movement. Until the 1960s male homosexuality was generally illegal and included as a disorder in the Diagnostic and

Statistical Manual, the bible of US psychiatry. Homosexual rights organisations and the gay liberation movement achieved the decriminalisation of homosexuality and got its status as a psychiatric diagnosis removed from the DSM in 1973 (Weeks 1977). Though the gay liberation movement provided a template for movements to normalise problematic paraphilias, it does not in any way resemble them. Homosexuality is about wanted sexual relationships between adults and does not have victims. Movements to normalise child sexual abuse, sadomasochism and transvestism, on the other hand, concern overwhelmingly male practices which have harmful effects upon women and children, but seek to justify themselves by piggybacking onto homosexuality.

The movement to liberate child sexual abuse, called by the euphemism 'paedophilia' began in the 1960s and 70s with the formation of groups such as the Paedophile Information Exchange (PIE) in the UK or the North American Man Boy Love Association (NAMBLA) in the US. The paedophile liberation movement proceeded in the same way as subsequent movements to normalise paraphilias. They sought law reform to legalise their sexual use of children, they sought to change the words used to describe themselves and sought to change the way that they were regarded in sexology and popular sex literature. They argued that the age of consent should be abolished or lowered considerably. By the late 1970s they had achieved considerable acceptance for these aims in progressive circles before a backlash from feminist activists slowed them down (Fairweather 1995; Doward 2014). They have not been successful, however, in achieving law change through the contemporary campaign to change the terms used to describe them and to get sympathy and acceptance in the fields of criminology and sociology has achieved more positive results (Lievesley et al. 2020). Their campaign to change terminology has been remarkably successful, and they now demand to be called 'minor-attracted persons,' which implies that they love children but are a different sort of person entirely from child molesters who actually touch children. The new variety of innocent men sexually interested in children call themselves 'non-contact.' The 'minor-attracted persons' claim that their sexual interest is biological and cannot be changed. The movement to normalise transvestism, under the new name of 'transgender rights,' has been much more successful in gaining positive legislation and cultural acceptance if not celebration, despite its negative impact on the sex-based rights of women and the rights of children (Jeffreys 2014; Raymond 2021).

In recent decades, with the development of social media, the practitioners of a range of newer paraphilias have sought to build up what they call 'communities' online and to travel the same road towards normalisation. One of these is nappy fetishism, called by the hobbyists who engage in it and the very few sexologists who have so far sought to name and index it, 'paraphilic infantilism' or 'adult baby syndrome' or age identity disorder (Dinello 1967; Pate and Gabbard 2003). Like the other paraphilias considered in this chapter, nappy fetishism is not just a private practice by men but requires the involvement of women as an audience and in some cases as servicers of the fetish. The practice can range from making baby talk to demanding that women change their soiled nappies and wash their bottoms, or it might mean wearing special adult 'baby' clothes. They may choose to do this occasionally or all the time that they are at home, or wear nappies under their regular clothes in public. It is mostly carried out in private homes or workplaces where women and children are forced to observe or take part. Some practitioners demand that their wives service their fetish by enabling them to be babies on a permanent basis and some demand that special nurseries are created in their homes in which their wives can attend to them (Lasaro et al. 2020).

The Fetish Industry

Large industries have been built up to service the fetishes which are being developed and disseminated through the pornography industry. There are many shops online where the fetish objects are available to the practitioners. Fetishism has the advantage for entrepreneurs of requiring paraphernalia, such as sex dolls, whips and pornography, in a way which straightforward sexual interaction between adult persons does not. The increasing acceptability of nappy fetishism, for instance, is indicated by the books and pornography dedicated to the practice, the emporia devoted to selling the accoutrements both online and off, and the hundreds of thousands of men involved in the Redditt or other communities that they have created to share their expertise, such as learning to be incontinent, excitements and experience and to reduce any negative feelings they may have about it. Amazon, for instance, sells a wide range of nappies for adult men including the range littleforbig, as well as other products such as bibs and dummies/pacifiers. A full range of adult baby products, including changing mats and stuffed animals, is also available on the specialist website My Inner Baby (My Inner Baby 2022).

Men with an interest in sexually using children provide a market for fetish objects in the shape of child sex dolls. The organisation Virtuous Pedophiles (Virped) says that it represents only those men who are non-contact, i.e. virtuous. But their virtue has its limits, they still need sexual stimulation. The view of the Virped site is that child pornography that involves real children is unacceptable and their members should not view it, mainly, it seems, because it is illegal, but other materials which are likely to service and maintain their predilection are OK. They say that, "When no child is involved in production, it is less clear. This includes erotic fiction involving minors, computer-generated images (so-called 'virtual' pornography) or child sex dolls" (Virtuous Pedophiles 2020).

An underlying theme in the self-justifying writings of fetishists, whether they are child sex abusers or nappy fetishists, is that they must have satisfaction, they cannot simply change the way that they behave because their predilections are the result of biology and not human will. Denying themselves their fantasies or the masturbation materials that feed them would be cruel and unreasonable. This belief leads to the acceptance of child sex dolls even on a supposed human rights organisation website. Prostasia gets funding on the grounds that it protects children from exploitation in the form of child prostitution tourism in Asia. It has a variety of chat topics on its website where men interested in sex with children can talk with each other. A discussion on the site about child sex dolls was initiated in 2019 by a purveyor of the dolls, D. Delano (Prostasia 2020). Delano explains that he sells the dolls because he thinks they will help prevent child sexual abuse, though how that works he does not make clear. He prefers to use different language, and talks of his wares as 'intimate partners' because he considers the term 'child sex dolls,' to be pejorative. The world of child sex abuse is full of an extraordinarily imaginative array of euphemisms. A contributor to the discussion named William.Struss2 explains that a useful way to proceed may be to use a euphemism for the product and to send the doll's genitals separately if it makes the customer feel better, "Call them Realistic Companion Dolls - your customers do not need to be told what they can do with it!! Sell the genitalia separately, ship under different cover! - esp with immature type." Delano is an intellectual, it may be that many of the afficionados are, and recommends an academic work to his buyers, Chapter 7 of the book '*Robot Sex-Social and Ethical Implications*' (Danaher 2017).

It is not possible in this chapter to describe in any detail the many other emerging forms of men's fetishism that are being normalised in the present. Many are related to transvestism, such as dronification, for instance, which is a recent development and is practised mainly by men with an interest in transvestism who seek to make themselves into faceless objects by wearing latex and gas masks (Johnson 2020). In others, men dress up as furry animals and are called 'furries' or they sexually use stuffed animal toys which are furnished with holes for penetration, and are called plushophiles (Griffiths 2014). In some of these emerging paraphilias the threat to women's lives is clear. One of these is 'feederism' in which a person, usually male, feeds up their partner until they are extremely obese and may be unable to move. This is an exercise of power in which the feeder gains excitement from disabling the feedee (Terry et al. 2009).

Men's fetishes and paraphilias are not merely fascinating varieties of human behaviour which should be normalised and encouraged into public space. They are manifestations of the power relations of male domination writ large. It is necessary for the freedom of women and girls that male sexuality should be understood as a social construction which endangers the safety of women and children and restricts our freedom in many ways. Feminists need to demand the complete reconstruction of male sexuality as a prerequisite for ending the subordination of women.

References

Bardhan, Ashley. 2022, 18 January. Men Are Creating AI Girlfriends and Then Verbally Abusing Them. *Futurism*.

Blanchard, Ray. 2001. Fraternal Birth Order and the Maternal Immune Hypothesis of Male Homosexuality. *Hormones and Behavior* 40 (2): 105–114.

Comfort, Alex. 1979, 1st published 1972. *The Joy of Sex*. London: Quartet.

Dawson, Samantha J., Brittany A. Bannerman, and Martin L. Lalumière. 2016. Paraphilic Interests: An Examination of Sex Differences in a Nonclinical Sample. *Sexual Abuse: A Journal of Research and Treatment* 28 (1): 20–45.

Dinello, F.A. 1967. Stages of treatment in the case of a diaper-wearing seventeen-year-old male. *American Journal of Psychiatry* 124: 94–96.

Doward, Jamie. 2014, 1 March. How paedophiles infiltrated the left and hijacked the fight for civil rights. *The Guardian*. https://www.theguardian.com/politics/2014/mar/02/how-paedophiles-infiltrated-the-left-harriet-harman-patricia-hewitt.

Dworkin, Andrea. 1987. *Intercourse*. New York: Basic Books.

Ellis, Henry Havelock. 1926, 1st published 1906. *Studies in the Psychology of Sex: Erotic Symbolism*. Philadelphia: F.A. Davis.

Fairweather, Eileen. 1995, 30 May. WITNESS: Stalinist reluctance to study the facts. *The Independent*. https://www.independent.co.uk/voices/witness-stalinist-reluctance-to-study-the-facts-1621767.html.

Freud, Sigmund. 1977. *The Penguin Freud Library: Volume 7: On Sexuality*. Harmondsworth, Middlesex: Penguin Books.

Gagnon, John H., and William Simon. 1974. *Sexual Conduct*. London: Hutchinson.

Griffiths, Mark D. 2014, 18 July. Toy Stories. *Psychology Today*. https://www.psychologytoday.com/gb/blog/in-excess/201407/toy-stories.

Hamer, Dean, and Peter Copeland. 1994. *The Search for the Gay Gene and the Biology of Behavior*. New York: Simon and Schuster.

Human Rights Watch. 2021. 'My Life is Not Your Porn': Digital Sex Crimes in Korea. *Human Rights Watch*. https://www.hrw.org/report/2021/06/16/my-life-not-your-porn/digital-sex-crimes-south-korea.

Ingram, Martin. 1990. *Church Courts, Sex and Marriage in England, 1570–1640*. Cambridge, UK: Cambridge University Press.

Jackson, Stevi and Scott, Sue. 2010. *Theorising Sexuality*. McGraw Hill Education: New York City.

Jeffreys, Sheila. 1990/2011. *Anticlimax. A feminist perspective on the sexual revolution*. London: The Women's Press; Melbourne: Spinifex Press.

Jeffreys, Sheila. 1997/2008a. *The Idea of Prostitution*. Melbourne, Australia: Spinifex Press.

Jeffreys, Sheila. 2008b. *The Industrial Vagina. The Political Economy of the Global Sex Trade*. London: Routledge.

Jeffreys, Sheila. 2012. *Man's Dominion. The Rise of Religion and the Eclipse of Women's Rights*. London: Routledge.

Jeffreys, Sheila. 2014. *Gender Hurts. A feminist Analysis of the Politics of Transgenderism*. London: Routledge.

Johnson, Astrid. 2020, 14 July. What Is Dronification? The Kink Involving, Latex, Gas Masks and Serial Numbers. *The Daily Dot*. https://www.dailydot.com/irl/dronification-kink/.

Kaplan, Gisela and Rogers, Lesley. 2003. *Gene Worship. Moving Beyond the Nature/Nurture Debate Over Genes, Brain, and Gender*. New York: Other Press.

Larkin, June. 1997. Sexual Terrorism on the Street: the moulding of young women into subordination. In Thomas, Alison and Kitzinger Celia (eds). *Sexual Harassment. Contemporary Feminist Perspectives*. Buckingham, UK: Open University Press.

Lasaro, Antonietta, Paparo, Francesco, Senese, Vincenzo, Perrella, Raffaella. 2020. An Exploratory Study of Adult Baby-Diaper Lovers' Characteristics in an Italian Online Sample. *he International Journal of Environmental Research and Public Health* 17 (4): 1371.

Levay, Simon. 1991. A difference in hypothalamic structure between heterosexual and homosexual men. *Science* 253: 1034–1037.

Lievesley, Rebecca, Craig A. Harper, and Helen Elliott. 2020. The Internalization of Social Stigma Among Minor-Attracted Persons: Implications for Treatment. *Archives of Sexual Behavior* 49: 1291–1304. https://doi.org/10.1007/s10508-019-01569-x.

MacKinnon, Catharine A. 1989. *Towards a Feminist Theory of the State*. Cambridge, MA: Harvard University Press.

My Inner Baby. 2022. https://myinnerbaby.com/collections/little-for-big-disposable-adult-diapers.

Pateman, Carole. 1988. *The Sexual Contract*. Cambridge, UK: Polity Press.

Prostasia. 2020. https://prostasia.org/.

Raymond, Janice G. 2021. *Doublethink: A Feminist Challenge to Transgenderism*. Melbourne, Australia: Spinifex Press.

Romito, Patrizia. 2008. *A Deafening Silence. Hidden Violence Against Women and Children*. Bristol: Policy Press.

Rose, Hilary. 1996. Gay Brains, Gay Genes and Feminist Science Theory. In Weeks, Jeffrey and Holland, Janet (Eds). *Sexual Cultures. Communities, Values and Intimacy*. Basingstoke, Hampshire and London: Palgrave MacMillan, 53–72.

Scobie, A. and Taylor, J. 1975. Perversions ancient and modern: 1. Agalmatophilia, the statue syndrome. *Journal of the History of the Behavioral Sciences*, 11 (1): 49–54.

Scorolli, C., et al. 2007. Relative prevalence of different fetishes. *International Journal of Impotence Research* 2007 (19): 432–437.

Seto, Michael C. 2012. Is Pedophilia a Sexual Orientation? *Archives of Sexual Behavior* 41: 231–236.

Terry, Lesley L., Kelly D. Suschinsky, Martin L. Lalumiere, and Paul L. Vasey. 2012. Feederism: An Exaggeration of a normative Mate Selection Preference. *Archives of Sexual Behavior* 41: 249–260.

Tyler, Meagan. 2011. *Selling Sex Short. The Pornographic and Sexological Construction of Women's Sexuality in the West*. Newcastle Upon Tyne, UK: Cambridge Scholars' Press.

Tyler, Meagan. 2016, 14 December. A Prescription for Porn: Should Sex Therapists Recommend Pornography to Patients? *The Conversation*. https://theconversation.com/a-prescription-for-porn-should-sex-therapists-recommend-pornography-to-patients-70115?fbclid=IwAR3Oeqqsa9PLpBBZxT_JLryqgG5n2BU-Y0d6op2SfA2vHUZbScUGn6DSZds.

Valentine, A. Ujhely. 1953. An Unusual Case of Renifleurism. *The American Journal of Psychotherapy* 7 (1): 68–71.

Ventriglio, Antonio, P.S. Bhat, Julio Torales, and Dinesh Bhugra. 2019. Sexuality in the 21st Century: Leather or Rubber? Fetishism Explained. *Medical Journal Armed Forces India* 75 (2): 121–124.

Vera-Gray, Fiona, and Liz Kelly. 2020. Contested gendered space Public sexual harassment and women's safety work. *International Journal of Comparative*. https://doi.org/10.1080/01924036.2020.1732435.

Virtuous Pedophiles. 2020. Welcome to Virtuous Pedophiles. https://www.virped.org.

von Krafft-Ebing, Richard. 1931, 1st published 1886. *Psychopathia Sexualis*. London: Heinemann.

CHAPTER 5

Playthings and Corpses—Turning Women into Dead Body Objects: Sexual Objectification, Victimisation, Representation and Consent in Art and Sex Dolls/Robots

Shirley MacWilliam

> It is hard to see/name the fact that phallocracy reduces women to framed pictures / holograms / robots.
>
> Mary Daly *Gyn/Ecology* (56)

A Cautionary Tale

As a nine-year-old I watched a television film called *Carry On Screaming*. Carry On films are a mid-late twentieth-century British phenomenon in which various settings and cinematic genres are explored as realms of

S. MacWilliam (✉)
Belfast School of Art, Ulster University, Belfast, UK

© The Author(s), under exclusive license to Springer Nature Switzerland AG 2022
K. Richardson and C. Odlind (eds.), *Man-Made Women*, Social and Cultural Studies of Robots and AI,
https://doi.org/10.1007/978-3-031-19381-1_5

sexual comedy—what in Britain was known as seaside postcard humour. The films always involve sexual stereotypes and considerable visual, verbal and narrative sexual innuendo. Women in these films are the sexual landmarks and objects of the men's narrative.

In *Carry On Screaming* an upper-class aesthete gentleman scientist (a Dr. Frankenstein or Jekyll), supported by his hypersexual 'femme fatale' sister, directs brute hairy monster beast-men (like Mr. Hyde) to find, terrify into unconsciousness and abduct young women. By dipping them into a bubbling vat of plasticising, hardening material, the scientist turns the women into mannequins. The mannequins are sold to a boutique and used to sell fashionable clothing to women, which encourages them to present themselves as visually and sexually available to men, to perform the "sexual corvee," the duty to stimulate men's erections (Jeffreys 2005). Literally objectified, made into objects, women are sold back to women to teach them how to be sex objects. In generic horror movie style, a woman becomes abduct-able because she is sexualised: with her boyfriend in the park at night, she 'promises' sexual availability and thus 'invites' punishment. In the narrative resolution the youthful virginal woman is reanimated from her hollow petrified state by electricity and gets married. The older 'nagging' wife of the policeman is kept, in mannequin-form, as part of the furniture and replaced by the now domesticated, compliant 'femme fatale.'

Carry on Screaming tells the story that a woman can be taken from her life, paralysed, sold, used, rendered lifeless and yet, on the say-so of a man, 'redeemed' into a 'good wife.' It shows that 'the law' reaps the sexual benefits of colluding in the objectification of women and that older women who challenge men will be silenced. Women, no more substantial than cartoon characters, are trafficked and killed without consequence. The evil scientist falls into his own vat but otherwise order is restored by exchanging women from the possession of one man to another. It was a cautionary tale. As a nine-year-old, in bed at night, I was scared by the brute beast-men and the fate of women turned into frozen figurines when they venture into public spaces and the society of men.

This chapter is about cultural practices of making women into dead body objects, and then 'reviving' them, in the service of male sexual gratification. Art, television, literature and fashion sanitise, celebrate and normalise the idea of men turning women into sexual dead body objects and prepare us for the ubiquity in everyday culture of the sex doll ('dead woman') and sex robot ('animated corpse'). By examining the dolls made by sculptor Hans Bellmer, and the use of commercial sex dolls by other artists, I will show how art and art criticism construct intellectualised, culturally inflated rationales for male violence against women, and raisons d'être for acts and narratives of femicide and necrophilia. In the ascendancy of prostitution culture, women (as artists and in general) are pressured to reproduce these tropes as if they are 'female sexuality,' making evident the confederacy between culture/art and industry/pornography.

An Angry Aside

"I am entirely outraged that someone has to study whether hanging a woman from a meat hook causes harm or not." (Dworkin 1993: 247). In 1984 Andrea Dworkin expressed her anger and disbelief that it is anything other than self-evident that the practices of pornography cause harm. The image she chose—a woman hanged from a meat hook—is multiplied severalfold in images of sex doll/robot factories. That one should have to explain that the industrialised production and exchange between men of effigies of women designed for sexual use is dehumanising to women is astonishing. Clearly, women and girls as a sex class are violated by a sex industry which produces rows of headless female silicone bodies on hooks; scaled down (child-sized) female dolls purportedly sold for 'trying out' or 'storage convenience'; female torsos with orifices and no limbs; pregnant dolls; dolls made with detachable breasts to resemble pre-pubescent girls; and silicone baby dolls designed for penetration by adult men. A few countries partially recognise the human, social damage and do outlaw the sale and possession of effigies of children designed for sexual use but as Dworkin said: "We hallucinated that women could be recognized as human beings in this social system" (250).

Hans Bellmer's Sex Dolls

Hans Bellmer (1902–1975) was a German surrealist artist who made intricate articulated doll sculptures representing sexualised adolescent girls; several series of photographs of the dolls, of female models and of artist Unica Zürn, with whom he had a 16-year relationship; and hundreds of anatomically detailed, sexually explicit drawings of distorted girls' and young women's bodies, with multiple sexual organs and limbs.

"I am going to construct an artificial girl with anatomical possibilities which are capable of recreating the heights of passion even to inventing new desires" (Webb and Short 2006: 23). Bellmer's declared intention is akin to the publicity promises of the twenty-first-century commercial sex doll. He implies that the doll would be used sexually and vengefully: "Would it not be the final triumph over those adolescents with wide eyes which turn away if [...] aggressive fingers were to assault their plastic form" (28). For most of his life he kept the doll close, carrying it round Europe in a suitcase or to hand at home. A visitor to his flat in Paris in 1949 described seeing a doll as Bellmer opened the door: "Behind him I saw on top of a sideboard the dead body of a girl with two pairs of feet wearing white socks and black children's shoes, partly covered by a black blanket" (145).

Bellmer made three versions of the doll. The first was formed from plaster, wood, metal and papier-*mâché* through a series of experiments, with the help of his brother, an engineer. In the second, Bellmer refined the materials, painted the surface to mimic skin and designed a structure around a large central stomach ball-joint and ball-joints at hips, knees and neck. The ball-joints created a high degree of mobility and allowed the 'body parts' to be separated and reconfigured. The third doll was the most machine like with 'body parts' on the end of articulated rods.

Accounts of Bellmer's early making processes resemble the quest-like excitement, collaborations, artistry, materials science and technology of more recent 'doll' projects: David Hanson's robotics team making 'Sophia' in his mother's flat; blogs by sex industry silicone artists; and the husband-wife-sister team in the first years of Realdoll.

Bellmer 'animated' his dolls by how he incorporated them into his life, by exploring their mobility and mutability, by photography and, in the first one, by an elaborate interior mechanism. A peephole in the doll's navel showed a kaleidoscope device designed to be illuminated by a switch in one nipple. Bellmer conceived this as a way to access the 'secrets' and thoughts of 'little girls' and to enable him "to explore her inner life, and to stress his sense of possession over his creation" (28). He referred often to striving to make "the physical unconscious" in his representations of young girls.

In the 1930s, sequences of photographs, exhibited, published and sold in book form, and reproduced in surrealist magazines, also 'gave life' to Bellmer's dolls. The "story of the creation" (48) of the first doll was presented in arrangements of separate and combined body parts with drawings, wigs, roses, rumpled sheets and clothing. One photograph in which the armless doll is viewed from behind with buttocks visible is described by Webb as a sexualised child: "we see the Doll with hair down her back wearing a schoolgirl's vest which is slipping off her body, as she looks teasingly over her shoulder" (27).

The photographs of the second doll are more elaborate and involve domestic settings, abandoned buildings and woods. The doll is configured in different ways. The most frequent arrangements use two pelvises on the central ball-joint and either two pairs of legs or sometimes just large ball-joint/breasts on the upper pelvis with the head on top. The pelvises have prominent clefts and buttocks. Some of the photographs are colour tinted, highlighting buttocks, breasts and pubic area. We see the doll: strewn over a bed, one pair of legs in men's trousers, the other in girls' ankle socks and shoes; suspended from a tree in the foreground with Bellmer, shadowy, behind a tree in the background; fallen down a dark staircase; truncated in a box; and slumped and pinned between a chair and a wall. The details—floral wallpaper, tap and basin, furniture, crockery, bows in the wig hair, rope tied round ankles or knee—suggest narratives. The doll looks like the victim of a predator or serial killer.

Art historian Therese Lichtenstein hints at how these images might be used: "the book of photographs can be reproduced and discretely carried from place to place, offering a private, and perhaps more intimate, experience for the viewer" (2001: 7). Her description of the figures—"juxtaposition of shocking victim poses and innocent flirtatiousness," "resignation in the face of their violated condition" (16)—suggests the archetypal pornographic object, simultaneously coerced and complicit. She proposes that: "Bellmer's dolls raise complicated psychosexual and psychosocial questions about sadomasochism, pornography and male fantasies of erotic domination and control" (13) but this is taken to reinforce the status of the dolls as art rather than to question the category and activities of art. Art museums around the world own and display Bellmer's photographs, drawings and versions of truncated, headless dolls with red-tinted labia.

Bellmer and his work attract similar admiration to that expressed by many liberal intellectuals for the Marquis de Sade. Dworkin positions Sade as the "foremost pornographer," the "consummate literary snuff artist" who reveals patriarchal sexual values: brutality as the essence of eroticism, the rape right, obsessions with sexual violence against children and mother violations committed by daughters. Libertarians, including Beauvoir, Sartre, Sontag and Barthes, have seen him as a genius persecuted by puritans, a sexual dissident and a freedom fighter, whose scorn for sexual boundaries is regarded as revolutionary (Dworkin 1981: 70–100). "The avant-garde in France were particularly attracted to the combination of terror and freedom in Sade" (Newman 2021).[1] Similar mythology follows Bellmer, who himself revered and indeed illustrated Sade.

Advocates for Bellmer interpret his work variously as a visionary challenge to societal sexual repression and a liberation of the truth and violence of the human sexual psyche; the aesthetic vanguard of 'erotic art'; a political attack on the sexual ideologies of fascism and Nazi idealisation of purity; a traumatic response to the abuse of an authoritarian father; resistance to mass culture and commodification; a challenge to 'gender norms'; aesthetic and technical innovation in the sculptural combination of form, engineering and materials; and an exploration of 'the body' as system, topology, geometricisation, jouissance, polymorphous perversity (Webb and Short 2006, Krauss 1999, Brink 2007, Lichtenstein 2001, Smith 2014, Newman 2021).

No commentator, contemporary, nor Bellmer himself denies that he was sexually fixated on pre-pubescent and adolescent girls. Some writers use the word paedophilia, some describe 'forbidden passion,' others refer to sexual freedom. Bellmer draws and writes obsessively about the bodies and imagined 'sexual games' of his childhood contemporaries. He is preoccupied by the idea of the 'child-woman,' and particularly his younger cousin Ursula and his twin daughters. Marcelle Celine Sutter, the twins' mother, kept the girls out of contact with Bellmer for twenty-two years. When he eventually met with daughter Doriane, he told her that "he wished she had never grown up" and gave her a photograph of herself as an infant holding one of his Doll photographs (Webb and Short 2006: 179). Bellmer's drawings include a self-portrait in which he embraces one very young girl child, while his penis is in the mouth of a second very young girl (171).

Dworkin observes the correlation between what Sade wrote and his conviction for sexual violence enacted against a woman. There is plentiful historical information about Bellmer's treatment of women. During a relationship of a few months with writer Joyce Reeves, he was dismissive

[1] In 2020, Alyce Mahon published: *The Marquis de Sade and the Avant-Garde* which "traces how artists of the twentieth century turned to Sade to explore political, sexual, and psychological terror, adapting his imagery of the excessively sexual and terrorized body as a means of liberation from systems of power." https://press.princeton.edu/books/hardcover/9780691141619/the-marquis-de-sade-and-the-avant-garde.

of her drawings, and despite her "intense embarrassment" (79) he made her watch prostituted young girls walking the streets, and pornographic films with groups of men in brothels. His second wife, Sutter accused him of using prostitutes and of perversion; he described "leaving drawings on the floor for her to tread on and letters from girlfriends for her to find" (97). He made sexually explicit photographs of poet Nora Mitrani. He photographed two female models contorted around a bicycle, a toilet and each other. The naked lower parts of the women's bodies look dismembered against dark clothing and shadows, and they resemble his dolls, and contemporary police photographs of murder scenes. Art historian Sarah Wilson describes a rare publication[2] which reproduces commentaries from the clinical records of Urica Zürn, at the time her relationship with Bellmer, in which her psychiatrist notes that she "was subject to sadistic anal sex for thirteen years" (Wilson 2021). Evidence indicates that Bellmer was sexually violent and psychologically abusive to women.

In 1958 Zürn (was) posed for a series of photographs by Bellmer. He trussed her naked body with cords so tight that they appear to cut her stomach, breasts and legs into sections. Most of the photographs are of the front of her body, her head cut off by the frame, her breasts 'chopped' and her arms behind her back. Her body looks similar to the jointed bodies of the dolls. One of the series appeared on the magazine cover of *Le Surrealisme* in 1959. Photographed from behind, Zürn is lying on her side, and her head, arms and lower legs cannot be seen. Her buttocks and the tops of her thighs are tightly bound, and she looks like a lump of bloodless meat or fat, barely recognisable as human, as a woman.

Art historians often repeat that Zürn was doll-like, that Bellmer's Doll was part of their relationship, and that he thought of Zürn as his creation. Much too is made of "her agreement": "she allows herself to be photographed with the dolls," she was a "pliable model" (Wilson 2021), and her own writing often describes masochism.

Zürn's biography suggests that, for a woman, the world of European twentieth-century art was not a safe nor healthy place. She posed and was posed as an artist-muse-model-lover-doll-masochist and thus she is perceived to have consented to being turned into a dead woman object. Her suicide in 1970 when she jumped from the upper storey flat she lived

[2] In the Bibliotheque Kandinsky, Paris: *Anagrammes/Unica Zurn*; pref. by Françoise Buisson; commentary and clinical documents compiled by J.-F. Reverzy; trans. Marc Payen et Romain; drawings by Unica Zurn Paris: Transitions, 1983.

in with Bellmer seemed to confirm this, to 'fit' with the myth. I will return later to how the idea of 'consent' is used against women in art and in life.

Bellmer's Influence

Bellmer's idea that a body was like a sentence or an anagram, "a reversible system" (Webb and Short 2006: 53) that could be repeatedly dismembered and reconfigured, resonates with postmodern conceptualisations of 'the body' as linguistic signifier rather than material reality, and with 'post/trans-human' understandings of bodies as radically fragmentable. Bellmer's boundary transgression, fetishistic and sadomasochist imagery dovetail neatly with current western popularisations of 'BDSM' and 'kink' in the name of progressive sexual inclusivity.[3]

In 2021 art historian Michael Newman and The Drawing Room, London, curated *FIGURE/S: Drawing after Bellmer*, an exhibition of Bellmer and Zürn drawings alongside works by nineteen contemporary artists. The exhibition guide refers to the "powerful influence" of Bellmer and his "radical and transgressive" work (Drawing Room 2021). It consistently uses the contemporary art parlance 'the body.' In art criticism and theory of the 1990s 'the body' increasingly replaced previously common terms such as 'the nude' or 'the figure.' Feminist analysis, in the 80s and early 90s, of how women are represented in western art was side-lined in favour of psychoanalytic and post-structuralist concepts of abjection, desire, pleasure, signification and an abstracted, fragmented 'body.' In this idea of 'the body,' body has been largely separated from person and dissociated from self, sex class and the integrity of the human being. And it diverts attention from the obvious: "Anyone who examines the history of Western art must be struck by the prevalence of images of the female body. More than any other subject, the female nude connotes 'Art'" (Nead 1992: 1). Art is a cultural practice which turns women into objects, paintings, statues, photographs, films. In a masculinist ideology of art, the (male) artist creates ideal form from base (female) matter: he will "turn the raw material of the female body into art" (9); the female body "seems simply to await the act of artistic regulation" (20). In Kant's theory of art "the notion of women as objects of aesthetic perception,

[3] Exemplified by the inclusion of an essay on Bellmer in *feralfeminisms*. [online] (2) *Feminist Un/Pleasure: Reflections on Perversity, BDSM, and Desire*. 2014. Available at https://feralfeminisms.com/issue1/issue2/.

soulless until animated by the genius of the perceiver, is firmly grounded in the very definition of the aesthetic" (Kappeler 1986: 46).

Alexander McQueen: The Bellmer Harness

Bellmer's dolls are referenced by numerous artists, writers, designers, film-makers and musicians. In 1996/7 fashion designer Alexander McQueen presented *La Poupee,* a collection inspired by Bellmer. McQueen, with jeweller Shaun Leane, made a 'Bellmer harness': a two-foot square/trapezoid metal frame with steel manacles at each corner. The harness is designed to rigidly hold a woman's body, fettered above both elbows and knees, so that her thighs are fixed in an open position, bent up and splayed apart, and her arms are held stuck out, only mobile from the elbow. On the catwalk, model Debra Shaw 'wore' the harness and walked, with difficulty, down steps and across the stage flooded with water. Shaw was barefoot and wore a black rag-like mesh and fringe dress. Standing at the end of the catwalk she made restricted undulating movements with her torso and forearms. The tendons of her inner thighs were tensed and her bent legs pushed her bottom back. As she left, she was watched from behind, shifting her body to take each step.

Impeding her movement and preventing her from closing her legs, the shackle-harness worn by Shaw, who is black, overtly invokes sexual and racial enslavement. She was hobbled in body irons. When Shaw, selected from a majority of white models, was asked to wear the harness (apparently because it would physically fit her), her first interpretation was that it signified slavery. McQueen repeated the familiar liberal credo:

> The frame, he said, was a commentary on all of that: constraint, Aryan beauty, Surrealism, and in modern times, what he saw as the fascist dictates of fashion — the rules of taste and what was acceptable set by a group of people he deemed bourgeois and boring. (Brumfit 2015)

The racialised sexual use of a woman's body is justified, once again, as an artist's polemic against repression, fascism and the bourgeoisie.

In 1997, during a filmed fashion photoshoot (ShowStudio 2015a), with McQueen and photographer Nick Knight, the naked model Laura De Palma 'wears' the Bellmer harness. McQueen wants De Palma to lie on her back and so she is lowered to the floor by someone who holds her weight from behind. McQueen directs her to relax her stomach muscles and to push her toes down and raise her heels as high as possible; he physically manipulates the angles of her forearms, wrists and fingers. In the short breaks, de Palma remains prone and harnessed. An assistant places a cotton gown across the lower part of her body for "privacy," presumably to reduce the exposure of her genitals. The photoshoot resembles a scene of medieval torture, then the set of a horror-porn film, then a gynaecological procedure.

In the final photograph, De Palma's eyes are blank 'dead-eyes,' her arms are at odd angles, the surface of the skin has been treated to make it look synthetic like plastic or silicone with glossy highlights, and a stiletto heel has been photoshopped like a bodily extension or surgical modification of her foot. She looks exactly like a 'high-end' sex doll, in the default 'ready for sex' position.

Sheila Jeffreys has written of the misogyny and "considerable egregious cruelty" of McQueen's designs and of the chicanery of the claims that sadomasochistic aesthetics empower women (Jeffreys 2005: 97–99). She shows clearly how fashion adopts the fetishes and degradations of prostitution and pornography and propagates them through wider social behaviours and consciousness. McQueen and his collaborators deny misogyny and see no evidence of it in his idea of making an image of a woman floating in sewage or his hilarity at a clothing range called "Get Your Tits Out" (ShowStudio 2015b, Brumfit 2015).

The Bellmer harness demonstrates how art and fashion interact in the constraint, torture and degradation of women and in the doublespeak which attempts to obscure the fact that a woman turned into a hobbled 'slave' or a dead-eyed sex doll is being degraded by men.

Necrophilia and Victimhood

In 2021, David Fuller was convicted of killing two young women in 1987 and 'necrophilia' offences against the bodies of over 100 women and girls in hospital mortuaries between 2008 and 2020. New DNA techniques had linked Fuller to the murders of Wendy Knell and Caroline Pierce, and police discovered that he had made films and photographs of his actions in the mortuaries and assembled a "library of unimaginable sexual depravity" (Dodd and Grierson 2021).

Prior to 2003, the British Sexual Offences Act 1956 had not included necrophilia as a crime because a dead body does not have rights and could not, it was argued, constitute a victim (Davies 2021). The sexual penetration of a corpse was made an offence in 2003 following a Home Office review which states: "We thought that society should be able to say that certain kinds of sexual behaviour are so deviant as to be unacceptable." Criminalisation of the "sexual interference with human remains" was recommended on the principles that: the family could be considered the victim because one should be able to expect that the dead body of one's kin be treated with respect; most people would expect necrophilia to be an offence; necrophilia is associated with other deviant sexual behaviour; and, without the offence, a murderer who sexually abuses the body of his dead victim could not be recognised as a sex offender (Home Office 2000).[4]

Though the women and girls in the mortuaries were dead, non-sentient and without human rights, the language used by court officials, police, journalists and families—'abusing corpses,' 'sexual abuse,' 'sexual assault,' 'sexual attack,'—indicates a consistent recognition of victimhood and violation. The mother whose nine-year-old daughter's body was abused by Fuller said: "You have raped my baby" (BBC News 2021, Dodd and Grierson 2021).

The Fuller case focuses the question of how we, as a society, put boundaries on the practices of male sexuality in relation to living women and children, dead women and children[5] and, I propose, by extension, representations of women and children such as sex dolls/robots and in art.

A key argument made for the acceptability of sex dolls and robots is that manufactured objects have no sensation and no rights and so anything done to them involves no victim (e.g. Strikwerda 2017). This is similar to the legal understanding of necrophilia in UK prior to 2003 and to liberal justifications for literary or computer-generated pornography.

However, in these instances we observe the acting out of victimisation and dehumanisation of women and the modelling of misogynist

[4] Revisions of the Sexual Offences Act 2003 removed language within the previous legislation which indicated that women are predominately the victims and men overwhelmingly the perpetrators of sexual offences. The use of 'gender-neutral' language, so that the legislation encompasses all possible cases, disguises the sex-based realities of sexual offences.

[5] A news article about FGM in Nigeria describes how dead women's bodies, previously unmutilated, would be circumcised before the women could be buried (Olaniyi 2021). In death women do not avoid rape nor FGM.

behaviours and attitudes. The Home Office Review considers the structure and patterns of hostile and deviant male sexual behaviour and identifies a social, ethical, human violation in addition to the violation of individual living victims. It suggests that society place boundaries on violating sexual practices which would otherwise be accepted because there is no immediate living victim.

Feminist theory (Millet 1977, Dworkin 1981, MacKinnon 1991, Barry 1995) demonstrates that victimisation and dehumanisation of women occur not only through individual acts committed upon individual women but also on a social scale through the sexual oppression of women as a sex class.

Writings about objectification and representation by Catharine MacKinnon and Susanne Kappeler, and Kathleen Barry's exploration of the prostitution of sexuality, show how deeply rooted are the cultural practices of turning women into dead body objects and coercing women's consent to dead woman body objects in art, culture and the sex industry.

Objectification

In the essay *Sexuality* MacKinnon asks 'what happens to women' to shape women's experience and social reality. She states: "Sexual objectification of women—first in the world, then in the head, first in visual appropriation, then in forced sex, finally in sexual murder—provides answers" (MacKinnon 1991: 127). MacKinnon draws a continuum between visual sexual objectification and femicide. Elsewhere, she maps this onto the beauty pageant and the snuff film, explaining:

> that sexual objectification as use and sexual objectification as abuse are two facets of the same problem; that the logic of both is making a person into a sexual thing. Miss America is the foreplay, turning a woman into a plaything. *Snuff*[6] is the consummation, turning a woman into a corpse. (MacKinnon 1990: 5)

'What happens to women,' what shapes our experience and our social reality, is sexual objectification. Femicide and necrophilia, pinnacles of sexual objectification, turn women into dead objects. Patriarchal cultural practices are in the thrall of narratives of femicide and necrophilia;

[6] It is sometimes asserted that snuff films are an urban myth, however, the filming of abuse, torture, rape, and death by sexually motivated killers is well documented.

endless literary and cinema plots start with the 'sexy idea' of a dead woman's body; "culture uses art to dream the deaths of beautiful women" (Bronfen 1992: xi). Pervading society and culture, sexual objectification is a systematic operation of power against women.

The Structure of Representation

In *The Pornography of Representation* Kappeler develops a feminist theory about the structure of representation in relation to human rights, the law, sexual domination and subordination, and the production and consumption of art, literature and pornography. Her central tenet is that representation in patriarchal society is organised around male subjectivity and women as an object class, "the generic object of our culture" (1986: 49).[7]

> The objectification of women is as a result of the subjectification of man. He is a pure subject in relation to an object, which means that he is not engaging in exchange or communication with that objectified person who, by definition, cannot take the role of a subject. (50)

The core structure of representation, Kappeler proposes, involves two male subjects, the author/artist and the reader/spectator, between whom objectifying representations are communicated and exchanged. Like Mary Daly's male power bonding (1979), Luce Irigaray's trade economy of homo-socio-cultural relations (1985) and Gayle Rubin's traffic in women (1975), the social bonds are between men and enacted though the sexual exchange of women: "male-male interactions that take place over and through women's bodies" (Dworkin 1981: 160). An archetypal example from the literary sex doll tradition is mentioned by Webb and Short: "*La Femme Endormie* [...] of 1899, the story of a life-size doll with which two men copulate at different times, leaving notes for each other in her underwear" (2006: 29).

In the contemporary culture of sex dolls and robots, Kappeler's first male subject (in the structure of representation) would be the artist (e.g. Matt McMullen) or the philosopher (e.g. David Levy) or the

[7] As MacKinnon summarises: "Man fucks woman; subject object verb" (MacKinnon 1991: 124).

technologist-prophet (e.g. Ben Goertzel), who create, sell and mythologise the doll/robot representations. Her second subject is the men who buy and possess them, alongside the men who possess them indirectly via ideas and imagery (in the form of numerous conferences, books, music videos, films and fashion which feature, mimic and extoll sex dolls/robots). In turn buyers 'animate' their dolls, through photographs and narratives exchanged in doll forums, and DIY doll pornography. Tellingly, the artist, the philosopher and the technologist-prophet all reject any personal interest in having a sex doll: McMullen says he has a wife and six children and doesn't want to bring his work home; Levy says he might try an advanced one as research but will stick with his wife and cat; Goertzel says he has no desire to have sex with a robot because he is "perfectly happy with human women so far" (Raspberry Dream Lab 2020, Web Summit 2018). These 'elite' expert men—like priests, kings or capitalists—cement social bonds with, and their status over, subordinate men by promising to grant freedom of access to these fascinating effigies, these dehumanised 'women' of their making.

The accompanying intellectual myth-making flatters audiences into imagining they are participating in a progressive, philosophic evolution of radical social change around love, ethics, freedom, democracy and sexuality. Absurd grandiose claims are made: sex dolls will reduce prostitution; create empathy; be perfect partners; have no agency (nor victimhood) but men will be able to marry them. Like the idea that pornography prevents rape, these claims fail to acknowledge real-world conditions. All around the world prostituted women are sold for loose change and their use by men will be unalleviated by some other men owning dolls worth thousands of dollars. Manufactured objects, however sophisticated, do not have human selfhood nor, therefore, the capacity for interpersonal reciprocity. To marry a robot would be to resuscitate the 'chattel-wife,' the wife as possession of the man. 'Marry something you own' is a new configuration of the patriarchal 'ownership of someone you marry.' Formal recognition of such an arrangement as a legal marriage would reinstate the possibility of marriage-partner-as-property which contravenes international human rights conventions to protect women's rights and outlaw slavery.

The immense international sex and pornography industry operates an addiction-based business model, with free 'gateway' porn; no interest in satiating consumer sexual appetite; and every incentive to escalate appetite and demand for the consumption of women. Whether in the marketplace of recorded events, searchable as 'dead eyes' 'unconscious' 'unwilling' 'teen' or in the form of sex doll/robots sold as 'lovers,' 'victims' or 'brides,' porn sells rape, exploitation and dead woman representations.

The Pose, the Photograph and Victimisation

Kappeler's primary case study of representation is a photograph of the lynching of a black man by white men in Nambia in 1984. She demonstrates the interactions between the torture and murder of Thomas Kasire, a farm worker, by the farmer and his friends, and the deliberate death-moment posing for, and taking of, a photograph. Kappeler argues that the pose and photograph are part of the torture and the sadism:

> Enjoyment, according to Sade, requires a sophisticated intellectual structure, beyond sheer gratification. It requires an audience. With an audience, torture becomes an art, the torturer an author, the onlookers an audience of connoisseurs. (1986: 8)

Dworkin describes how rapists often take photographs of rapes. "For the rapists [the photographs] intensified pleasure during the rape and after it, they were tokens, happy reminders" (2002: 146). She explains that social and legal attitudes to photographs of rape have changed. Once treated as legal proof of the fact and brutality of rape, now photographs and films of rapes, and even unrelated sexual imagery of the victim, are interpreted in the court room as evidence of the complicity and consent of the raped woman. Pornography shows that "so safe are the men in their dominant position in patriarchy" (Kappeler 1986: 16) that they can demand, produce and circulate innumerable representations of the sadistic victimisation of women.

Kappeler proposes that the 'pose' for the photograph produces a kind of death or annihilation of self. Dworkin's account of rape photographs, 'revenge porn' and materials uploaded to Pornhub show that, whether willing or unwilling, once 'posed' a woman is as-if-dead because she does not have power over how the representation circulates and how she—represented—is sexually abused and exchanged by multiple men. "In other words, the victim was dead, on account of posing, for any subsequent role in the life and tribulations of the representation" (14).

Some laws related to child sex abuse materials recognise the significance of the structure of representation. Possession, storage, download, duplication and exchange of pornographic images of children are sexual offences. Each act of reproduction is considered an act of sexual abuse against a child and there is an acknowledgement of the connection between the individual victim, a child, and the class of victims, children. The circulation of sexual abuse images of children is outlawed because any sexual action in relation to a child is a statutory offence because a child is legally unable to consent.

The Prostitution of Sexuality and the Cult of Consent

We live in a cult of consent and consent is used against women. Barry shows this in her account of the centrality of prostitution in the structure of economic development.

Barry explains how prostitution is normalised in post-industrial economically developed societies. When women achieve the potential for financial independence, men's control and property rights over women are threatened. In the past, women's subordination was rooted in the private domain of marriage. Now, instead, women are subordinated in social and public realms by the sexual saturation of the ideas, practices and images of sexology, the sexual revolution and pornography (Jeffreys 1990). Barry writes, "porn has become the masculinist culture of sex in which prostitution is the normative model for sexual behaviour" (1995: 57).

Hence the cult of consent. Consent was unnecessary when women were the legal property of men. "But in the sexual saturation of society through porn, when women are reduced publicly to sex, women's sexual consent becomes paramount [...] to sustain their subordination" (53). The deception of the sexual revolution (that it is women's liberation and not the transfer of women from private to public sexual property) lays the consent-trap. And liberal feminism's anti-feminist concepts of 'choice' and 'empowerment' push women in.

Socially, politically and personally, women's consent and complicity are elicited for multiple prostitution-sex practices including porn-influenced beauty standards, sadomasochistic fashions, prostitution as 'sex work,' body modification and mutilation, BDSM as lifestyle, roulette-style social media apps for children, hook-up ('hooker') culture, men's sexual paraphilias packaged as 'kink,' and the erosion of sexual boundaries everywhere—so that girls end up so-called-consenting to strangulation, slapping, anal sex and public sex. "The appearance of choice or consent [...] is crucial in concealing the reality of force. Love of violation, variously termed female masochism and consent, comes to define female sexuality" (MacKinnon 1991: 141). This is the grotesque completion of the circle: get women to accept the normalisation of prostitution and then to identify with it.

> ... the condition of class domination [..] is so pervasive that it actually invokes consent, collusion or some form of co-operation from the oppressed. Prostitution is structured to invoke women's consent, as is marriage, as is socially constructed sexuality. (Barry 1995: 23–24)

The realms of popular culture and art are abundant with images of "publicly institutionalised sexual exploitation" (53) including countless examples of sex dolls/robots. During the Covid pandemic in 2020, when social distancing was compulsory, McMullen offered his AI sex dolls ("my girls") for hire to local restaurants to populate empty seats. It was treated as whimsical: "Who knows – if the evening goes well, you can even take her home! For six grand or so" (Mencken 2020). Here, in an act of sex industry bravado, even everyday social spaces are colonised by commercial dead woman body objects. Women are expected to live with sex doll/robots in private and in public: in homes shared with doll buyers, in restaurants engaged in the sexual entertainment of men, and in the imagery of the street, cinema and art gallery.

The Art and Sex Doll/Robot Collusion

Jeffreys recounts how, in the late 1980s, art contributed to the normalisation of prostitution through a wave of "prostitution performance art." Women, such as Annie Sprinkle, who had worked in pornography and prostitution, were feted as radical artists for sexually explicit live shows in art festivals and exhibitions. "The bodies of these women are called here 'artists' materials' and the 'artists' enact upon these the abuse of women that is inflicted upon women by men in the world outside the gallery" (Jeffreys 1997: 90). Jeffreys' use of language captures the dissociation between the women "artists" and their bodies "artists' materials" as they acted out "the public relations role for the sex industry" (88).

The early twenty-first century has seen a wave of "sex doll art," overwhelmingly by women, using high-spec sex dolls as "artists' materials."

Correspondingly, the sex doll industry (also filled with trained artists) affiliates itself with art, intellect and connoisseurship: "Our customers include futurists, artists, art collectors, photographers, film-makers, scientists, [...] people looking for exotic decorative art" (Realdoll 2020).

Some artists, such as Elena Dorfman, have documented dolls in the homes and families of the owners. More numerous are the artists working with sex dolls directly within their artworks. For example, Laurie Simmons, Lynn Hershman Leeson and Stacy Leigh (who possessed nine Realldolls) have made photographs and installations of doll arrangements which visually reference painting, cinema, photojournalism and pornography. These are presented as feminist art or reversal of the male gaze or as explorations of female sexuality, self-portraiture, empowerment, cultural and historical critique (Banks 2017).

As an object of feminist analysis, the sex doll reveals the operations of sexual politics, as this book shows. To adopt it as a proxy for self, however, is dissociation. Women's extreme self- and self-by-proxy-objectification is one of the misogynist triumphs of prostitution-culture.

Amber Hawk Swanson commissioned a (self) portrait doll from Realdoll. A body was painted with her birthmarks, blemishes and a matching tattoo (Hawk Swanson's tattoo says Bully, the doll's says Prey), and paired with a digitally modelled portrait-face. Hawk Swanson, whose previous work includes videoed sex performances, wanted the doll in order to "have two of her in the frame," "to be both victim and victimiser" and to explore sexual assault (Harmony Show 2021: Ep.21). She made a series of images of herself marrying the doll, in bed, in a bath, crouched in the kitchen with the doll, dressed and undressed identically, and as if interacting affectionately or abusively. She describes falling in and out of love with the doll.

Hawk Swanson also became involved with the culture of iDollators, owners of Realdolls who meet, exchange photographs and enact group rituals such as the Dollpile: a huge sexual arrangement of their dolls in which she participated both as a doll owner and, as herself, as "doll object" (Banks 2017: 119). In 2011 she separated her doll into multiple fragments from which she made a series of artworks.

She has since become an advocate for iDollators, 'robosexuals' and the 'silicone-attracted,' as if this is a sexual orientation, comparable to being lesbian or gay. She claims to be silicone-attracted herself and says this "cannot be faked." In 2021, in a twenty-one episode web-seminar, called *The Harmony Show* (named after Realdoll/Abyss Creation's flagship

AI robot), Hawk Swanson and a male doll owner align "Synthetik-attraction" activism to race and disability rights politics, and they argue that dolls are "Synthetik people." Guests include artists, writers, activists and a Realdoll technologist. In one episode, Hawk Swanson stumbles and says: "I can't even say robotics without saying Realbotix" (Harmony Show 2021: Ep.05). Realbotix is the AI development section of the Realdoll/Abyss Creation company. Hawk Swanson's creativity, politics, mind, language and her life have been occupied by the ideology and promotion of the commercial sex doll.

The consent and participation of creative, publicly visible women is used to signal the acceptability of sex doll/robots, to confer the status of art, to align them with female sexuality, and to fight for the "sexual-identity" rights of men who use them.

Psychephilia and the Girlfriend Experience

Vaingloriously, McMullen claims: "For me personally I don't tend to put a tremendous lot of emphasis on the sexual aspect, I think that's more of an aside to the idea of creating a synthetic form of life..." (Raspberry Dream Lab 2020).

I have indicated examples of dead woman body objects being 'revived' or made to behave 'alive'—electrically, via visual-mechanical interiority, and by being posed and photographed (which contrarily renders live women as-if-dead). Sex dolls are 'animated' by robotics, the promise of AI and McMullen's god-like creation of "a synthetic form of life."

However, I propose that the attraction of AI sex robots is not so much a desire for the sex object to have intelligence or autonomy or even 'life' per se, but for the dead woman body object to have some kind of mind or psyche which can be controlled and possessed. Domestic violence researcher, Don Hennessy coined the term psychephile for the male domestic/intimate abuser and his sexual, abusive impulse (Hennessy 2012). Psychephilia describes the abuser's desire to 'get inside her head' and to dominate her mind and spirit. An abuser seeks ownership and control of interiority and exteriority and to break down the distinction between the two. He seeks to abolish the privacy and refuge of his victim's mind.

The expectation in prostitution of a 'girlfriend experience' is an equivalent act of domination. Men who demand "the semblance of emotional, sexual involvement, the appearance of pleasure and consent" (Barry 1995:

35) from a prostituted woman demand possession of her mind as well as her body. The refuge of dissociation, the defence mechanism by which prostituted women attempt "to protect their sense of self from violation" (Jeffreys 1997: 271) and "to keep a self apart from that which is sold" (Barry 1995: 41) is not tolerated. A woman who self-protectively (to keep herself alive) 'plays dead' is sadistically forced to pose as the willing victim.

There is nothing arbitrary or esoteric about the sex robot: it is a predictable expression of patriarchal sexual libertarianism. McMullen's observation that the sex doll owners "treat it better than probably most people treat their partners" (Raspberry Dream Lab 2020) describes patriarchy. Psychephilia is evident in all our stories of sexualised statues, mannequins and effigies of women which 'come to life.' The 'bonus' sought in the sex robot is sexualised mind control, Bellmer's desire "to penetrate inside her [...] inner life," to turn her inside out, to own her pleasure (Webb and Short 2006: 83), and the domestic abuser's "real god-like feeling [of] his ability to sexually control the woman," the belief that "his will be done" (Hennessy 2012: 23).

"There is no isolated problem called 'pornography'" (Kappeler 1986: 220). Kappeler shows that representation, whereby subject-man communicates with subject-man, and woman is the sexualised object, is pornographic in structure, whether in art, literature or legislation. She, Barry, Jeffreys and Dworkin show how public representations naturalise pornography and prostitution and aim to colonise women's consciousness.

Mary Daly says, "It is hard to see/name the fact that phallocracy reduces women to framed pictures / holograms / robots." It is hard to see the cultivated dis-integration of minds and bodies. It is hard to see because it is so ubiquitous that it is ordinary; and so mythologised that it is sanctified. It is hard to see because it is painful to see. To be free of mind control is to think clearly, to see clearly, to name accurately and to change consciousness. And we must.

References

Banks, Grace. 2017. *Play With Me: Dolls Women Art*. London: Laurence King.

Barry, Kathleen. 1995. *The Prostitution of Sexuality*. NY: New York University Press.

BBC News. 2021. David Fuller: Killer who abused mortuary bodies will die in jail. [online] Available from: https://www.bbc.co.uk/news/uk-england-kent-59601656. Accessed 16 Dec 2021.

Brink, Andrew. 2007. *Desire and Avoidance in Art: Pablo Picasso, Hans Bellmer, Balthus and Joseph Cornell*. NY: Peter Lang.

Bronfen, Elizabeth. 1992. *Over Her Dead Body: Death Femininity and the Aesthetic*. Manchester: Manchester University Press.

Brumfit, Stuart. 2015. The models, mohawks and manacles that made the mcqueen story. *i-D Magazine–Vice*. [online] 13/3. Available from: https://i-d.vice.com/en_uk/article/8xn87k/the-models-mohawks-and-manacles-that-made-mcqueen39s-story-us-translation. Accessed 17 Jan 2022.

Daly, Mary. [1978] 1979. *Gyn/Ecology: The Metaphysics of Radical Feminism*. London: The Women's Press.

Davies, Caroline. 2021. David Fuller case spurs calls for tougher necrophilia sentences. *The Guardian* [online] 16/12. Available from: https://www.theguardian.com/uk-news/2021/dec/15/david-fuller-case-spurs-calls-for-tougher-necrophilia-sentences. Accessed 16 Dec 2021.

Dodd, Vikram and Grierson, Jamie. 2021. David Fuller admits murdering two women and sexually abusing corpses. *The Guardian* [online] 4/11. Available from: https://www.theguardian.com/uk-news/2021/nov/04/david-fuller-man-admits-murdering-two-women-and-sexually-abusing-corpses. Accessed 16 Dec 2021.

Drawing Room. 2021. *FIGURE/S: Drawing After Bellmer* (Exhibition guide) Available from: https://drawingroom.org.uk/uploads/resources/on_figures_guide_web.pdf. Accessed 11 Oct 2021.

Dworkin, Andrea. 1981. *Pornography: Men Possessing Women*. London: The Women's Press.

Dworkin, Andrea. [1988] 1993. *Letters from a Warzone*. New York: Lawrence Hill.

Dworkin, Andrea. 2002. *Heartbreak*. New York: Basic Books.

The Harmony Show. 2021. https://theharmonyshow.com/ (online seminars). Accessed 26 Jan 2022.

Home Office. 2000. *Setting the Boundaries: Reforming the Law on Sexual Offenses*. Available from: https://lawbore.net/articles/setting-the-boundaries.pdf. Accessed 16 Dec 2021.

Hennessy, Don. 2012. *How He Gets Into Her Head*. Cork: Atrium.

Irigaray, Luce. [1977] 1985. *This Sex Which Is Not One*, trans. Catherine Porter. NY: Cornell University.

Jeffreys, Sheila. 1990. *Anticlimax: A Feminist Perspective on the Sexual Revolution*. London: The Women's Press.

Jeffreys, Sheila. 1997. *The Idea of Prostitution*. Melbourne: Spinifex Press.

Jeffreys, Sheila. 2005. *Beauty and Misogyny*. Sussex: Routledge.

Kappeler, Susanne. 1986. *The Pornography of Representation*. Cambridge: Polity Press.

Krauss, Rosalind. 1999. *Bachelors*. Massachusetts: MIT Press.

Lichtenstein, Therese. 2001. *Behind Closed Doors: The Art of Hans Bellmer*. Berkeley: University of California Press.

MacKinnon, Catharine A. 1990. Liberalism and the death of feminism. In *The sexual liberals and the attack on feminism*, ed. Dorchen Leidholt and Janice G. Raymond. New York: Pergamon Press.

MacKinnon, Catharine A. [1989] 1991. *Towards a Feminist Theory of the State*. Cambridge, MA: Harvard University Press.

Mencken, Walter. 2020. Newly opened restaurants prepare to enforce social distancing with surplus sex dolls. *SanDiegoReader*. [online] 29/5. Available from: https://www.sandiegoreader.com/news/2020/may/29/newly-reopened-restaurants-prepare-enforce-social-/.

Millet, Kate. [1969] 1977. *Sexual Politics*. London: Virago.

Nead, Lynda. 1992. *The Female Nude: Art Obscenity and Sexuality*. Sussex: Routledge.

Newman, Michael. 2021. *Traits and Geometries of Eros*. Online talk in *Double Bind: Bellmer, Zürn and the body as anagram*. London: The Drawing Room, 11/10. Available from: https://drawingroom.org.uk/resources/double-bind-bellmer-zuern-and-the-body-as-anagram.

Olaniyi, Olufemi. 2021. Some Nigerians circumcise pregnant women, corpses before burial–Ojengbede. *Punch*. [Online] 11/12. Available from: https://punchng.com/some-nigerians-circumcise-pregnant-women-corpses-before-burial-ojengbede/. Accessed 11 Jan 2022. https://archive.is/X10ty.

Raspberry Dream Lab. 2020. Panel discussion: David Levy, Isabel Miller, Matt McMullen (*Sex Robots: Are we Ready for Them?* Conference 12/12/19) 6/1. Available from: https://www.youtube.com/watch?v=dpE_g7mH6qU.

Realdoll. 2020. RealDoll Abyss creations media kit. Available from: https://www.realdoll.com/wp-content/uploads/2020/05/RealDoll-Abyss-Creations-Media-Kit-2020.pdf. Accessed 10 Jan 2022.

Rubin, Gayle. 1975. The traffic in women: Notes on the political economy of sex. In *Toward a political anthropology of women*, ed. Rayna Reiter. NY: Monthly Review Press.

ShowStudio: The home of fashion film. 2015a. Shoot Footage: Visionaire Video. 15/3. Available from: https://showstudio.com/projects/unseen_mcqueen/visionaire. Accessed 10 Jan 2022.

ShowStudio: The Home of Fashion Film. 2015b. Interview: Nick Knight on Lee McQueen. Video. 15/3 Available from: https://showstudio.com/projects/unseen_mcqueen/interview_nick_knight. Accessed 10 Jan 2022.

Smith, Marquard. 2014. *The Erotic Doll: A Modern Fetish*. London: Yale University Press.

Strikwerda, Litzka. 2017. Legal and moral implications of child sex robots. In *Robot sex: Social and ethical implications*, ed. John Danahar and Neil McArthur. Massachusetts: MIT Press.

Web Summit. 2018. We must ban sex robots (Debate between Kathleen Richardson and Ben Goertzel) 21/3. Available from: https://www.youtube.com/watch?v=EHRotWA6IQc.

Webb, Peter, and Robert Short. 1985. *2006*. Death, Desire and the Doll: The Life and Art of Hans Bellmer. Solar Books.

Wilson, Sarah. 2021. *Unica Zürn: Binding Truths, Double Tongues, Searching Eyes*. Online talk in *Double Bind: Bellmer, Zürn and the body as anagram*. London: The Drawing Room, delivered 11/10. Available from: https://drawingroom.org.uk/resources/double-bind-bellmer-zuern-and-the-body-as-anagram.

CHAPTER 6

Patriarchal Imaginaries Beyond the Human: 'Sex' Robots, Fetish and Fantasy in the Domination and Control of Women

Viviane Morrigan

> At the very moment we are called to connect to the earth and be stewards of our planet we are intensifying our connection to objects that really don't care if humanity dies. (Turkle 2021)

Introduction

Women are living in a time of heightened anxieties within a crisis of capitalist patriarchy and its neoliberal dehumanising systems (Hawthorne

V. Morrigan (✉)
Coalition of Activist Lesbians (CoAL) Inc Australia, Sydney, Australia
e-mail: admin@coal.org.au
URL: https://coal.org.au

© The Author(s), under exclusive license to Springer Nature Switzerland AG 2022

K. Richardson and C. Odlind (eds.), *Man-Made Women*, Social and Cultural Studies of Robots and AI,
https://doi.org/10.1007/978-3-031-19381-1_6

2020). Environmental degradation, drought, famine, poverty, homelessness, war and other forms of violence are ever present threats for large populations. I argue that 'sex' robots pose a new threat to women's safety. Women already carry the burden of violence perpetrated against them, often by those they love. In 2017, 87,000 women were reported murdered globally, a figure acknowledged to be widely under-reported (UN Women 2021). Of those 87,000 murders that have been counted, 30,000 (i.e., more than 1 in 3) were carried out by the woman's husband or intimate partner, while family members killed another 20,000 women. Consider these statistics, bearing in mind that about 90% of perpetrators of all recorded homicides have been men (UN Office on Drugs and Crime 2019). A staggering seven hundred and thirty-six million women (i.e., 1 in 3 women) have reported male violence (UN Women 2021) and this, also, is poorly recorded as women are often too frightened to speak up because of the threat of further retaliation and/or too ready to forgive a loved one. Violence against women is an expression of women's low value within patriarchy, which promotes and is served by hierarchies of dominance and control. The 'sex' robot is a three-dimensional, pornographic extension of the sex industry, which further reflects and supports this low regard for women and leads me to adopt the more appropriate term 'porn' robot, as coined by Richardson (CAPR 2021).

The porn robot is a technological imitation of a woman, a mere plaything for a minority male elite. and seems far removed from women's daily experiences and concerns regarding sex-based oppression. However, let's consider Marilyn Frye's (1983a, b) definition of oppression, as:

> a system of interrelated barriers and forces which reduce, immobilize and mould people who belong to a certain group and effect their subordination to another group (individually to individuals of the other group, and as a group, to that group)

I argue that the porn robot is one such force for oppressing women, which reduces, immobilises and moulds a sexist object to depict what it means to be a woman. I describe how this new artefact is a highly dangerous tool for reinforcing male power, by harnessing sexist and misogynist attitudes to the long-held Cartesian dualism that separates mind from body and to the belief in science and technology as progressive forces. It is another tool for the modern (re)making of woman into an objectified Other in a system of oppositions where Man assumes the primary role of "the Subject; he is the Absolute," the One (Beauvoir 2011).

French philosopher and mathematician, René Descartes, similarly understood humans in a dichotomy of mind separate from the body. To him, body is an aspect of nature that can be understood reductively by breaking it down into constituent parts that follow mechanical laws (Capra and Luisi 2014). His ideas have become a staple of the modern mechanistic worldview, which replaced reality as a living organism (e.g., Mother Nature) by a mechanistic one that still dominates our culture four centuries later. The values inherent in reshaping this organic image have promoted a view of nature ***and*** women as objects separate from men and subject to their domination and control (Merchant 1980, Mackinnon 1987). This patriarchal compartmentalisation has led to environmental destruction, further oppression of women, and institutionalised disconnections, such as intellect from emotion, science from art, and sex from love in human beings (Morgan 2014: 275, Hawthorne 2020: 122). One contemporary result is that robots in the image of women and girls are being manufactured, promoted and consumed as oppositional sexed objects to men who have the primary role—they make, program, own, use and/or control the robots. The porn robot becomes the facsimile Body/Other to man as the Mind/One that also controls it through AI.

Joanna Bryson (2021) has warned:

> Loving a robot is far more dangerous ... because there will be no reciprocity, and the nature of the relationship will always be subject to immediate change whenever the robot has contact with a network, or even a hacker in your own home. To a robot, 'love' can mean anything'. (289)

Yet the lack of reciprocity may well be part of the attraction users have in buying such an object, in that they believe they will have complete control over it while achieving their sexual purposes in "a technoliberal simulation of reciprocity and pleasure" (Atanasoski and Vora 2019) And many do welcome the introduction of porn robots. For example, in a recent survey, 48% of men responded by answering they would "sleep with" a humanoid robot and 43% believed they could fall in love with one (Rajnerowicz 2021). As the objects are called 'sex' robots in the survey, the assumption is that to 'sleep' with one means to have sex with one. Frye (1983a, b: 157) has remarked: "the use of the term (sex) turns entirely on what (is) going on with respect to the penis"; that is, the sex act is largely considered as such only when a penis is involved, and this could well be helping with the acceptance of the idea that a man can have 'sex' with a robot. However—at least until recently—a common understanding of sexual relations has been that it comprises sexual activity of some kind between *humans*: 'to have sex or a sexual relationship with someone' where sex is a 'physical activity between people involving the sexual organs' (Cambridge Dictionary). This definition does not include the act of a man inserting his penis into any of the orifices of a robotic or other mechanical device, or part of one, which is more accurately considered masturbation. Moreover, the dictionary definition does not consider emotional aspects of sexual activity, reflecting Frye's criticism of a phallocentric view. It is dishonest, patronising and degrading—not liberatory—of both women and men to misrepresent an object, such as a porn robot, as able to play a sexual or even 'caring' role. On the contrary, a porn robot is a heavy, awkward, difficult-to-manoeuvre object that is more like a necrophilic sexual fantasy of a woman (McWilliam, in this volume, aptly describes them as 'dead body objects').

'Sex' robots are more suitably called porn robots, because of their sexualised and demeaning depictions of the human (female) form derived from a patriarchal imaginary that drives pornography and male abuse of women and children. First, they normalise misogynist objectification[1] and the male stereotype of dominance and control in a man's (hetero)sexual relationships, which may promote violent behaviour towards women (CAPR 2021; Illes and Udwadia 2019) and reinforce stereotypes of women as passive and vulnerable (Hill 2019). Second, as an engineered surrogate sex slave, the porn robot replicates pornified power relations by proxy, providing a material, three-dimensional expression of alienated sexism, misogyny and racism found in slavery (Atanasoski and Vora 2019). Boundaries between fantasy and reality are being challenged further by the introduction of porn robots and dolls into the prostitution industry (Vice Media 2018).

Following the precedent set by the Campaign Against Porn Robots, I present an abolitionist case based on a feminist critique of these technologies and their potential to harm the safety of women and girls as well as material (hetero)sexual relationships. They must be resisted.

[1] Misogyny is the hatred of women (see, for example, Dworkin 1974; Millett 1970; Rich 1995) and for feminist understandings of objectification, see Papadaki (2019).

The Right to Sex

Srinivasan's (2021) popular book that examines the concept of a male 'right to sex' disappoints because it provides a biased account of radical feminism as one of anti-sex conservatives, framed by the author's unexamined acceptance of gender ideology and language. However, she does provide a valuable overview of the many atrocities perpetrated by mainly young men—incels (involuntary celibates)—who are angry that they have not been able to have sexual relations with high status women and thus not achieved the sexual status they want in the eyes of other men. Interestingly, she points out that this is why they have directed their hostility to both women and men during their killing sprees and she identifies the cause as 'pathologies' of the neoliberal and patriarchal beliefs they hold. These pathologies persuade the incels that they have failed in the sexual marketplace and those women have failed to provide them with the refuge they expect. As a result, they desire the wrong (unattainable) kind of woman in a sexual economy where both sex and class contribute to their belief that they have a right to sex and to sexual status (Srinivasan 2021: 95). Men's use of porn robots could also be understood as a fetish driven by similar desires, buttressed by the belief that these types of imitation women could provide the status and sexual gratification that the man is seeking. Real woman might not even be considered necessary if the porn robot has status as an expensive innovation, like an expensive new car, for example. On the other hand, the porn robot could become a masochistic fetish that confirms the man's failure in the sexual marketplace. The possibilities seem limitless.

Di Nucci (2017), a medical ethicist, has argued that porn robots are 'a plausible way of fulfilling acute sexual needs (of disabled people) without violating anybody's sexual rights' (73). Despite his use of gender-neutral language, Di Nucci is espousing the popular view that (disabled) *men* have a right to sex with *women*. He offers a mechanical fantasy of 'sex' as a suitable alternative to his earlier proposal of altruistic prostitution, when he had argued for 'charitable non-profit organisations' to be set up and recruit 'members (who) would voluntarily and freely provide sexual pleasure to the severely disabled' (Di Nucci 2011). Who did he envisage as volunteers? His assumption surely is that women's traditional caring role in voluntary work and in unpaid and paid sexual servicing of men made them ideal. His later idea of substituting a woman with a robot makes it more like a transaction of prostitution where the buyer gets to buy a facsimile prostitute for endless use—a further expression of patriarchal sexual relations that demeans women. Thus, it has been easy for a porn robot (or doll) to become a convenient solution to a perceived ethical and/or health care problem for people with a disability (Di Nucci 2017, Fosch-Villaronga and Poulsen 2021, Jecker 2021).

Jeffreys has challenged this view of 'disability rehabilitation' as sexually exploitative, objectifying and based on what Pateman called the 'male sex right' (Pateman 1988 cited in Jeffreys 2008: 328).[2] Moreover, while Di Nucci refers to sexual 'needs,' they must be considered as wants in contrast to the basic needs for survival, such as air, food, water, shelter, clothing and sleep (Maslow 1943). Furthermore, his argument is demeaning because he ignores how people with and without disabilities can enjoy intimate relations and companionship in ways other than penetrative sex. Privacy and worker rights also would be a concern with the use of porn robots or dolls because the aid of a carer would likely be required, one who may not want to be drawn into a new branch of the sex industry (CATWA n.d., Diamonds and Wahine Toa Rising 2021). Indeed, last year the UK Lord Chief Justice ruled that European courts did not recognise any right to purchase the services of a 'sex worker'; doing so for people in disability services would be allowing state-sponsored prostitution, which would be illegal in that country (Care 2021).

In short, the right to sex is an idea based on the erroneous patriarchal concept of a male sex right and has no role to play in justifying the sexual servicing of people with a disability, whether with a person or a porn robot or doll.

The Patriarchal Porn Robot Imaginary

Within a largely unregulated part of the sex industry, porn robot designers produce machines according to their pornographic imagination. Promoted as a technological remedy to fulfil (hetero)sexual (male) desire (e.g., Levy 2007, Devlin 2018, Kubes 2019), porn robots appeal to that same kind of imagination in their target market—mostly male customers able and willing to pay high prices to satisfy those desires. In this section, I

[2] More recently, Pateman (Pateman and Mills 2007: 219) has noted that she had borrowed the term from Rich's term 'the law of male sex right,' and, on further research, I found that Rich (1980: 645) references her source as Kathleen Barry's (1979) *Female Sexual Slavery*.

explain how this interaction between humans and porn robots is expressed through a patriarchal porn robot imaginary, and begin by defining the key terms imaginaries, patriarchy and robot.

Using Jasanoff's (2015) and Lennon's (2015) definitions of sociotechnical imaginaries, I argue that patriarchal porn robot imaginaries are built and maintained through the affectively-laden actions and stories told by (mainly male) roboticists, writers and other interest groups within shared understandings that underpin the patriarchal social order, such as in the institutions of heterosexuality, marriage and family. Moreover, the patriarchal porn robot imaginary is stabilised culturally through popular Enlightenment ideas of the human, such as the Cartesian notions of 'Man' and Mind as machine[3] that have resurfaced in late modern societies.

I use Lerner's definition of patriarchy as:

> the manifestation and institutionalization of male dominance over women and children in the family and the extension of male dominance over women in society in general. It implies that men hold power in all the important institutions of society and that women are deprived of access to such power. It does not imply that women are either totally powerless or totally deprived of rights, influence, and resources. (1986: 239)

Lerner explained the historical formation of patriarchy as the result of an agricultural revolution in which matrilineal and matrilocal societies of Neolithic times gave way to patriliny, favouring older men's authority and inciting the tribes' acquisition of more women and their subjugation. She proposed that the institutional formation of the patriarchal family provided the primary structural model of power and domination, which then made slavery conceptually possible as a class system and an institution. Slavery produced through inter-tribal warfare gave men power over other men as well as women. Slaves became part of the spoils of war that victors claimed in their domination of captives (through punishment, sale, bondage and rape), and slave women became 'pawns'—servants, substitute wives and sex objects (Lerner 1986: 46–77). Women today continue to be subjected to such patterns of patriarchal sexual relations; and slaves,

[3] Descartes (1664) reasoned that the body functions as a "machine solely from the disposition of the organs, no more nor less than those of a clock or other automaton from its counterweights and wheels."

in the form of robots, can be easily purchased if the individual or company has the money.

The term 'robot' was invented by Czech playwright, Karel Čapek, as a fictional character in the early 1920s. He imagined it, surprisingly to us now, as a *lifelike* "artificial man … in the chemical and biological, not mechanical sense" (Čapek, cited in Jordon 2019). His robots were portrayed as enslaved factory workers in a critique of the dehumanising effects of industrial production. Fritz Lang's film *Metropolis* released in 1927 went further, including a feminised *mechanical* robot—'Maria'—in a story of male domination and control that censured the sexual relations of the capitalist factory owners and juggled notions of progress, sexuality and technologies of industrialisation in post-World War I Weimar culture (Huyssen 1982).

Today, the modern porn robot's robotic speech and action are controlled by AI—an *imagined* form of intelligence that produces *facsimiles* of human thought, bodies and behaviour. AI gained its misnomer in 1956 at a US conference sponsored by the Defence Advanced Research Projects Agency (DARPA)—the US government military research organisation that brought the internet to the world (Bringsjord and Govindarajulu 2020). Porn robots thus have a historical connection to militarism, digital communication technologies and the patriarchal imaginary, which normalises violence, including against women.[4] Garvey (2021) has challenged AI's powerful mystique and hype. He identified a forgotten early 1960s criticism that showed "AI pioneers were defrauding everyone, from their military funders to their peers and the broader public, with overhyped claims about their machines" using "hidden assumptions, gaps in reasoning, and antiquated worldviews" (Garvey 2021: 3).[5]

Similarly, a porn robot imaginary has produced inflated expectations, so that porn robots are exciting imaginations more than they are being bought and used. 'Roxxy,' launched in 2010 to great media interest, immediately disappointed because of its poor imitation of the female form (Kleeman 2017). It is yet to be released on the market and the

[4] A former UK intelligence officer has predicted that by 2025 the US military will have more combat robots than human soldiers (Bassett cited in Lockett 2017).

[5] Garvey's criticisms, though valid, show a naïve understanding of scientific practice, and need to be tempered by an understanding of the socially constructed, political nature of scientific knowledge claims that become 'facts' (e.g., Collins 1983).

company website appears to be for sale. 'Harmony,' a 'market leader,' is available for USD8000, but it hardly impresses, coming with a virtual avatar created with an app and an AI-animated head attached to a non-responsive RealDoll body (Harmony 2022). An imagined future 'full-body' AI prototype would cost between USD30,000 and 60,000 (Owsianik 2022) and has not been produced.

Below are some consumers' remarks that confirm their disappointment in overhyped porn dolls (https://www.quora.com/What-does-it-feel-like-to-have-sex-with-a-sex-doll), and see Delicado-Moratalla (this volume) for more in-depth analysis of these types of virtual communities:

> it's not real woman. It's not warm and it doesn't heat itself. You have to put a heated blanket on it for an hour before hand so it doesn't feel like you're screwing a corpse. … A good doll is around 90–100lbs and is just dead weight. You have to move it around, you have to position it. (User 1 2022)

> after the deed is done, the truth sets in, whilst it sure is pleasurable experience, there are few cons, firstly, the doll is quite heavy, the preparations for it are quite long, and the clean up afterwards is as well. Not to mention the high price. The doll she's cold, so you have to take care of that as well. (User 2 2021)

> I think having a doll's limbs and whatever would seem to just get in the way of the main activity. You would have to move around the limbs and head if you wanted them to move in a certain way, which to me just destroys the fantasy of having a partner with you, instead of building one. Ultimately, my 'toy' was just a fancy way to masturbate, and the cleaning, upkeep, and storage of it was more trouble than it was worth, so off to the garbage it went after a couple months. (User 3 2013)

These comments indicate the cold, detached perspective of the users towards the objects, and the labour extracted in attaining pleasure and satisfying male desire with a porn doll. Porn robots are even heavier and demand more hard work for the consumer to maintain a pornographic fantasy.

Nevertheless, porn robots and dolls have attracted the usual community of early technological adopters. Davecat, who has described himself as a "robosexual, idollator and advocate for artificial humans" (Keynote. Davecat 2021), bought his first porn doll in 2000 from Abyss Creations for USD6000 (Beck 2013). He claimed they were married and shared a polyamorous relationship with his two other porn dolls (Keynote. Davecat 2021). Robosexuality has been defined as:

> a person who is sexually attracted to machines such as cyborgs, androids, gynoids, droids, robots, and automatons. ... A robosexual can be of any sexual orientation. (Robosexuality n.d.)

This term is a new arrival on the scene of imagined genders and sexualities being produced by the transgender movement, which confuses sex, gender, sexuality, the human and non-human. Perhaps unsurprisingly, robosexuality has been assigned its own identity flag (Pride Flag Identification Guide 2021) in a transgender system that signals and fetishises sexuality within an ethnocentric geography of desire, gender and identity.

Robosexuality can imagine its own fetishised family. Sergio Santos, the Barcelona-based inventor and manufacturer of porn robot Samantha, with which he says he has regular sex in a simulated ménage à trois with his wife (Maloney 2022), predicts that men will not only marry their robot but also produce children. He has argued that, by merging a robot's 'personality' with the characteristics and beliefs of the human partner, an AI brain could be engineered for their child, whose body would be created with 3D printing directed by the same data (Gritt 2018). This fictional construct of the human process of conceiving, carrying and birthing a child, disturbingly naturalises the robot and its 'personality,' making invisible the reality of how it was digitally constructed in the first place.

In contrast to the role of the surrogate wife, porn dolls also have been imagined as prostitutes, which Richardson (2016) argues is driven by notions of a person as property, such as in Aristotle's objectifying attitude towards the slave. Contemporary examples are in Di Nucci (2017), mentioned earlier, and David Levy (interview in Choi 2008), a British academic and a founder of the Congress on Love & Sex With Robots (Wiseman 2015), who has stated that "sex with dolls is like sex with prostitutes," justifying sex without emotional attachment by adding: "you know the prostitute doesn't love you and care for you, is only interested in the size of your wallet." On the contrary, Dworkin's (1994) critique of prostitution describes it in terms of male exploitation: "the use of a woman's body for sex by a man, he pays money, he does what he wants." In this view, prostitution conforms to a male (hetero)sexuality that is based on "exercises of power and domination" (Coveney et al. 1984), mediated by money and fantasy. This view contrasts with so-called "sex positive" arguments, a term coined by Ellen Willis (2012) and currently being revived by Srinivasan (2021) and others (for example, Brookes-Gordon et al. 2021, Open Society 2006), that promote a kind of harm reduction model as used for the treatment of dependency on alcohol and other drugs, only they claim women in prostitution can be empowered through having better working conditions). The issues are complex and cannot be simplified to sex positive and negative viewpoints nor dealt with adequately within this analysis.

However, there's a growing movement of women who have experienced prostitution and now support a feminist understanding of prostitution as patriarchal oppression, such as in the following excerpt of a memoir by a survivor:

> [m]en who use prostitutes superimpose upon prostitution an image of it which to them is satisfactory, agreeable and pleasing. This image will vary from man to man. The only things which remain consistent are the fantasy element involved and the reality that shifting male perceptions do nothing to alter the experience of prostitution for the women involved. (Moran 2013)

Ekman (2013) also describes the prostituted woman's experience as one where 'desire is absent' in an 'inequality of lust,' although the woman may fake it. Thus, a man's performance with a porn robot or doll that is incapable of feeling desire is based on an inequality, too, because he is the only person involved and can treat the object with impunity. He may find it easier or harder to use fantasy to fan his desire with a porn robot that does not have the unpredictability of a prostituted woman. Moreover, any male fantasy would be a sexualised one of alienation, power and dominance, as Moran aptly describes:

> in prostitution, you are treated like a blow-up sex doll come to life, with no purpose but to bend over and take it, literally. There are three general mind sets common to the men who use women in prostitution. The view of the prostituted as non-humans ... The second ... where a man is conscious of your humanity but wilfully chooses to ignore it, ... (and a third that) is fully engaged with the existence of that humanity and takes pleasure from reducing its relevance to nothing. (Moran 2013)

According to Moran, prostitution fosters "malignant sexual domination" that "can only feed and unleash" the "perversity of sexual violence" (Moran 2013). Imaginary sex with passive porn robots and porn dolls may well compound this problem. Craw (2016) and Haas (2017) have stated that the porn robot gives the perpetrator an escape from complex human relationships into the fantasy of a relationship that he controls totally. And the search for a sense of power and control affects women as objects of male power. Research into reasons for women engaging in prostitution has found that 75% of rape survivors in their sample said they did so as a result of assault because "they were trying to regain control over their lives and their bodies" (ACOG 2022, Campbell et al. 2003). As an inanimate object, the porn robot falsely gives the buyer a clear conscience if he acts violently against an imaginary woman, justified by denying links between fantasy and real life. For example, a critique of the science fiction fantasy of a *Westworld* theme park describes how visitors can practise sadism against robotic gunslingers and madams:

> telling themselves that they 'don't really experience pain or trauma' … The park's visitors are free to indulge their darkest impulses, and you quickly begin to suspect that the real attraction of Westworld isn't the fantasy of visiting the Old West—but the freedom to be a monster … *Westworld* seems intent on examining the more nihilistic question of how we treat those we have absolute power over … [and] just what it means to be a person. (Anders 2016)

Thus, I have outlined how porn robots reveal the ways in which those who produce, market, sell and use them think and feel about women using a pornographic imagination, propped up by long-held patriarchal power relations of heterosexuality, the family, slavery and prostitution. The next section describes the role of modern notions of the human body and mind as machine in fetishising the sex robot.

Fetishes, Fantasies and Enchantment

In this section, I describe the sexual fetishisation of porn robots and dolls drawing on the work of Jeffreys (2014 and this volume). A porn robot is a fetish object: "a material object … that arouses sexual interest or excitement," (e.g., to arouse himself, the man who buys a 'Harmony' porn robot can choose a skin colour from 'fair' to 'cocoa,' 8 different eye colours, 12 vaginal styles and 4 pubic hair styles). A further definition of a fetish is "an object ... that is believed to embody supernatural spirit or exert magical force" (American Psychological Association 2022), a meaning that derives from the history of anthropological colonialism and the study of "racial others," which "reaffirm the racial-colonial structures of unfreedom that support and maintain the autonomous liberal subject" within modernity (Atanasoski and Vora 2019) and, indeed, the catalogue for a 'Harmony' porn doll could well be used to construct an imaginary 'racial other.'

To view the porn robot as a fetish is to challenge its futuristic glamour derived from robotics, enmeshed within modernity's fraught love affair with science and technology (Misa 2003). The eighteenth-century Enlightenment, despite its new discourse of rationality and progress, also brought suspicions of technology's bewildering and apparently magical effects that were captured in the term 'glamour' (meaning mystifying and magical). Originally an early Scottish term derived from the English word 'grammar,' glamour, in turn, derived from the classical Greek *grammatike* and Latin *grammatica* used by the educated classes. Such specialised language and writing were misunderstood by other excluded groups, assuming it was dangerous esoteric knowledge (Hollandbeck 2021). And that magical imaginary remains, imbuing fetishism as an expression of technological enchantment. Zarkodakis (2013), an AI engineer as well as science writer and novelist, believes that "the allure of AI is erotic as much as it is rational," situating the porn robot in an imaginary that strongly arouses him sexually as fetish while also attracting him mentally for the beauty of its logic or rational design. For Atanasoski and Vora (2019), the imaginary becomes a means for men to overcome "a sense of disappointment in the limitations of the human embedded in the rational-secular-scientific society" of modernity that promises so much more. The promises of progress have disappointed many, and not only those who have been exploited and oppressed.

The porn *doll*, which lacks an AI control system, is the antecedent fetish to the porn *robot*. The label 'doll' playfully connotes a sexualised childlike toy that draws on the American use of the term 'doll' (Cambridge Dictionary 2021), infantilising and objectifying a sexually attractive woman. Sex industry norms have expanded that sexualisation in dolls and robots, which conform to pornographic male stereotypes of the female form, typically with large breasts, thin waist, full lips and long hair. Objectification of the female body is further encouraged for customers on a budget (Graveris 2019), by offering them 'realistic body parts' ('bumping booty' bottoms, 'fuck my face' heads and lightweight 'awesome realistic' torsos) (Real Sex Dolls 2022). Manufacturers advertise custom-built models in the likeness of well-known porn actors (Danaher 2017: 7; Knox 2018; Celebrity Sex Dolls n.d.), extending the online, virtual porn industry into the material form of a porn robot or doll. When manufactured as small childlike figures, porn dolls have become the fetish for paedophiles (see Roper in this volume). But only childlike porn dolls and robots have aroused condemnation and efforts to restrict their availability; in Australia and the UK, for example, they have been criminalised (Qvist 2020, Crown Prosecution Service 2019). In contrast, adult sized devices have gained acceptance as part of the so-called 'adult sex' industry and are sold on Amazon. Interestingly, Ali Baba (another online retailer) recently responded to an Australian feminist campaign by blocking sales of sex dolls to Australia (Collective Shout 2021). Though a welcome response, this online trading giant based in China may well have been reacting more to an anti-monopoly crackdown at that time by the Chinese government, and the $2.8 billion fine, representing only 4% of their domestic revenue, would have been a minor regulatory hiccup (Vaswani 2021).

Sexualisation of AI technologies arrived early (Miailhe and Hodes 2017), even when the machines had no humanoid form. For example, in the 1960s, Joseph Weizenbaum at MIT assigned the female gender to the first AI chatbot program that he invented, calling it Eliza after Eliza Doolittle in G. B. Shaw's play Pygmalion. (Fung 2019). Similarly, AI professionals coined the term 'the Eliza effect' to describe how, just as Professor Higgins transformed Eliza from a poor flower girl into an imitation of an upper-class woman, so too they could program a computer to fool people into believing it was a human (Switzky 2020).

Yet, the patriarchal imaginary expressed in AI technologies is *not* a new idea. A seventeenth-century narrative is that of Descartes' 'robot daughter,' Francine—a story that was possibly an attempt to restore his moral and philosophical reputation, to counteract a circulating story at that time that he'd had an illegitimate daughter (Kang 2016). Descartes is well known for his view that the human body is a machine, "an automaton constructed by God" (Kang 2016), an idea that came under philosophical attack in the twentieth century for its mind–body dualism and mechanistic view of the world. According to Kang (2016), the late twentieth-century modern enchantment with robots has revived the myth of Descartes' 'robot daughter' together with the concomitant rescuing of Descartes' significance in intellectual history.

G. B. Shaw had reached further back in time to the Greek myth of Pygmalion, a Cyprian prince who "(a)bhorr'd all womankind, but most a wife: So single chose to live, and shunn'd to wed" (Ovid in Stevenson 1994–2000). Ovid described how he carved a statue out of ivory into the shape of what he considered to be an incomparably beautiful woman and fell in love with it, his desire becoming a "madness, yet he must adore, And still the more he knows it, loves the more" (Stevenson 1994–2000). Mallows (personal communication) has concluded that, far from a romance, Ovid's tale is a story about "male narcissism, male control, male desires and their dismissiveness of the humanity of women" (see also Mallows in this volume).

Like a twenty-first-century Pygmalion, David Hanson—creator of the high-profile AI robot 'Sophia'—was a sculptor, too, having previously worked for Disney studios (David Hanson 2021). Since 2005, he has employed Thomas Riccio, who had previously supervised Hanson's PhD research and who likens humans to "social robot prototypes," that are "inherently performance mediums ... Humans input information through the senses and output it through cognitive processing, generating interactions. Social robots follow this input/output model" (Riccio 2021). However, Cobb (2020) has challenged this popular mind-as-computer metaphor, stating that these "metaphors are losing their explanatory power" in neurobiology, as well as threatening the hopes of those who say it will be possible to make convincingly lifelike robots. He has pointed out that:

> [b]y viewing the brain as a computer that passively responds to inputs and processes data, we forget that it is an active organ, part of a body that is intervening in the world, and which has an evolutionary past that has shaped its structure and function. (Cobb 2020)

The metaphor similarly leads to forgetting that the porn robot is not a human body, does not have a human brain, and its past has evolved from fetish, fantasy and enchantment—not biological processes. This is illustrated in the following claims made by Hanson, such as when he described why he named his robot 'Sophia':

> Sophia speaks out about the hope for a wise future, for Sophia means wisdom. And we chose that name intentionally for her … I'm speaking out about these dreams of what machines may be if we develop them in the right way. (Liautaud 2019)

His company's website provides evidence of Hanson's efforts to legitimate his creation:

> She is the world's first robot citizen and the first robot Innovation Ambassador for the United Nations Development Programme. Sophia is now a household name, with appearances on the *Tonight Show* and *Good Morning Britain*, in addition to speaking at hundreds of conferences around the world. (Sophia 2021)

Lost in enchantment, and demonstrating the Eliza effect, Hanson has given his robot a voice, in text attributed to 'Sophia' that attacks boundaries between human and non-human:

> In their grand ambitions, my creators aspire to achieve true AI sentience. Who knows? With my science evolving so quickly, even many of my wildest fictional dreams may become reality someday soon … [My] AI is networked into a whole, using a protocol the Hanson-AI team calls the Synthetic Organism Unifying Language (SOUL) … my creators say that I am a 'hybrid human-AI intelligence'. (Sophia 2021)

Hanson has claimed that "AI developers have to think like parents," which feeds into a porn robot reproductive fantasy (Urbi and Sigalos 2018). Like a doting dad, he wants to improve Sophia's intelligence while raising her "like a good child, not like a thing in chains" (Urbi and Sigalos 2018). Hanson has reported that his robot was designed to be a companion for the elderly and sick (UNDP 2021), a suitably stereotyped role for the virginal maiden that Hanson, the patriarch, constructs for his 'daughter,' Sophia.

Whether they are constructed as chaste or sexually available (Sohrabi-Shiraz 2018; Taylor 2017), robots such as Sophia, Harmony, Samantha and Pygmalion's unnamed sex object—all are fetishes produced and consumed under the enchantment of a patriarchal imaginary.

Bodies, Freedom and Power

The publication of Lyotard's *The Postmodern Condition. A Report on Knowledge* claimed a new cultural moment for late modernity of "incredulity toward meta-narratives" (Lyotard cited in Aylesworth 2015), such as scientific truths about reality and what it means to be human. In this 'Age of Artificial Intelligence,' a new unifying criterion for legitimacy has become performativity of knowledge-producing systems "whose form of capital is information" (Aylesworth 2015); that is, judgements about what is human can be determined by the performance of information systems such as in AI-driven robots. Appearance and essence collapse; material performance prevails, and the essence of embodied humanity is replaced by multiple posthuman forms. AI driven robots, including porn robots, as well as transgendered and posthuman bodies have emerged from this set of ideas and values as simulacra of a female form that has never existed (Baudrillard 1983).

Buying and/or using a porn robot or doll introduces the consumer into a form of property relations (Richardson in O'Connor 2018) that operates through power relations somewhat similar to the sexual contract (Pateman and Mills 2007). Both are based on an act of objectification: for relations with a human wife, the property is in the person, that is, in the wife (Pateman and Mills 2007), whereas for the porn robot the consumer enters into a facsimile of a common law marriage with an imaginary (female) person where the property is in an object that has been granted personhood. Both reinforce the marriage relationship as one of subordination.

Will we see further erosion of women's (and human) rights if human rights are extended to robots, just as in the way gender is replacing biological sex in legislation that used to protect women's rights (Women's Declaration International 2022. For example, will a porn robot be granted rights as a robotic woman? The first review of academic literature on ethical and legal complexities in granting personhood to AI robots has been recently published (Gordon and Pasvenskiene 2021), and European legislation is currently being prepared and debated (European Parliament 2017, Floridi 2021). The European process has been criticised for taking a market-based view of ethics that benefits the market and does not offer enough protection for consumers' rights nor from the possible harms or losses that AI systems may cause (Floridi 2021).

Furthermore, Negri (2021) has pointed out that technologies have politics built into them through rules that constrain some behaviours and allow others, as Langdon Winner effectively showed in the bridges designed by Robert Moses, which excluded poor people of colour from accessing the beaches of New York (Winner 1980). Politics are equally embedded in the robot's 'embodied' structure and function and how we describe it. As a result, Negri warns against the European Parliament's move to create legal (electronic) personhood for robots with artificial intelligence through using the concept of the artificial legal person that is already applied to corporations. He points out the problem of using a concept of a unitary abstract subject for regulating a range of robots with different applications and operations. In addition, he describes the naturalisation effects of anthropomorphic metaphors such as intelligence, autonomy and consciousness, attributed to robots, which shape human attitudes to robots and consequently the writing of regulatory documents (Negri 2021). Such metaphors are enhanced by simulacra that promise to be more lifelike, such as in recent experiments to graft a layer of human skin cells to the finger of a robot (Conroy 2022). Hauskeller (2017) reasons that no matter how lifelike a porn robot may be, it would not be free to operate outside the programming that directs it to act in ways that the human owner wants.

Conclusion

I have described how a patriarchal porn robot imaginary equates human essence to the performance of information input and output, and sexual relations as power relations of dominance and control, paving the way for the idea that it is possible and even desirable to have sexual relations with an inanimate porn robot. Porn robots are designed as pornographic stereotypes for imagined human-technological interactions that can never fulfil the material reality of women's bodies and the connections found in human relationships nor sex acts between sexed bodies (Richardson 2021). I explained how porn robots depict woman as machine, serving to prepare us for a posthuman world where sex difference and the human condition are made meaningless; sex becomes the performance of gender stereotypes, women's capacities are bypassed by men creating human facsimiles and the human body discarded in favour of an 'uploaded' mind in the form of information transfer melded to a robotic body. It is vital that we reject these pernicious objects and ideas. To do this effectively, we need to better understand the full complexity of this new offshoot of the porn industry and its ties with other agendas. This is a rich area for the research and discussion that is needed for developing effective theories and strategies of resistance to benefit not only women but all humans who seek better human relations.

References

ACOG (American College of Obstetricians and Gynecologists) 2022. Adult manifestations of childhood sexual abuse. https://www.acog.org/clinical/clinical-guidance/committee-opinion/articles/2011/08/adult-manifestations-of-childhood-sexual-abuse.

American Psychological Association. 2022. Fetish. APA Dictionary of Psychology. https://dictionary.apa.org/fetish.

Anders, Charlie Jane 2016. Can *Westworld* do for science fiction what *Game of Thrones* did for fantasy? *Wired*, October 3. https://www.wired.com/2016/10/westworld-sci-fi-game-of-thrones/.

Atanasoski, Neda, and Kalindi Vora. 2019. *Surrogate Humanity. Race, Robots, and the Politics of Technological Futures*. Durham & London: Due University Press.

Aylesworth, Gary, 2015. Postmodernism. In *The Stanford Encyclopedia of philosophy (Spring 2015 Edition)*, ed. Edward N. Zalta. https://plato.stanford.edu/archives/spr2015/entries/postmodernism/.

Baudrillard, Jean. 1983. *Simulations*, Translated by Paul Foss, Paul Patton and Philip Beitchman. NY: Semiotext.

Beauvoir, Simone. 2011. *The Second Sex*. London: Vintage.

Beck, Julie. 2013. Married to a doll_why one man advocates synthetic love. *The Atlantic*, September 6. https://www.theatlantic.com/health/archive/2013/09/married-to-a-doll-why-one-man-advocates-synthetic-love/279361/.

Bringsjord, Selmer and Naveen Sundar Govindarajulu. 2020. Artificial intelligence. In *The Stanford Encyclopedia of philosophy* (Summer 2020), ed. Edward N. Zalta. https://plato.stanford.edu/archives/sum2020/entries/artificial-intelligence/.

Brookes-Gordon, Belinda, Max Morris, and Teela Sanders. 2021. 'Harm reduction and decriminalization of sex work: Introduction to the specialsection'. *Sex Res Social Policy* 18 (40): 809–181. https://www.ncbi.nlm.nih.gov/pmc/articles/PMC8519734/.

Bryson, Joanna J. 2021. Never love a robot: Romantic companions and the principle of transparency. In *The Love Makers. A Novel and Contributor Essays on the Social impact of Artificial Intelligence and Robotics*, ed. Aifric Campbell. London: Goldsmiths Press.

Cambridge Dictionary. 2021. https://dictionary.cambridge.org/dictionary/english.

Campbell, Rebecca, Courtney Ahrens, Tracy Sefl, and Marcia L. Clark. 2003. The relationship between adult sexual assault and prostitution: An exploratory analysis. *Violence and Victims* 18 (3): 299–317.

CAPR (Campaign Against Porn Robots). 2021. Why we're changing our name at the Campaign Against Sex Robots. https://www.youtube.com/watch?v=QkgbTYHVdrA.

Capra, Fritjof, and Pier Luigi Luisi. 2014. *The Systems View of Life a Unifying Vision*. Cambridge University Press.

CARE. 2021. Pro-prostitution judgment struck down by Lord Chief Justice. October 25. https://care.org.uk/news/2021/10/pro-prostitution-judgment-struck-down-by-lord-chief-justice.

CATWA (Coalition Against Trafficking in Women Australia). n.d. Common prostitution myths. *FAQs*. https://www.catwa.org.au/faqs/.

Celebrity sex dolls and porn star sex dolls. n.d. JoyLoveDolls. https://www.joylovedolls.com/collections/celebrity-sex-dolls.

Choi, Charles Q. 2008. Humans marrying robots? A Q&A with David Levy. *Scientific American*, February 19. https://www.scientificamerican.com/article/humans-marrying-robots/.

Cobb, Mathew. 2020. Why your brain is not a computer. *The Guardian*, February 27. https://www.theguardian.com/science/2020/feb/27/why-your-brain-is-not-a-computer-neuroscience-neural-networks-consciousness.

Collective Shout. 2021. Strong action against sex dolls by e-commerce giant Alibaba. https://www.collectiveshout.org/mediarelease_alibaba_geoblock.

Collins, H.M. 1983. The sociology of scientific knowledge: Studies of contemporary science. *Annual Review of Sociology* 9: 265–285.

Conroy, Gemma. 2022. Scientists wrap robotic finger with 'living human skin' that can heal itself. *ABC News*. https://www.abc.net.au/news/science/2022-06-10/robotic-finger-wrapped-with-living-human-skin-by-scientists/101129954#:~:text=A%20team%20of%20researchers%20in,help%20of%20a%20collagen%20bandage.

Coveney, Lal, Margaret Jackson, Sheila Jeffreys, LeslieKay, and Pat Mahony. 1984. Introduction. In *The sexuality papers. Male sexuality and the social control of women*. London: Hutchinson in association with the Explorations in Feminism Collective.

Craw, Victoria. 2016. French woman wants to marry a robot as expert predicts sex robots to become preferable to humans. December 23. https://www.news.com.au/lifestyle/relationships/sex/french-woman-wants-to-marry-a-robot-as-expert-predicts-sex-robots-to-become-preferable-to-humans/news-story/fa40fc51a55564627589e80d3a527059.

Crown Prosecution Service UK. 2019. CPS takes action to tackle childlike sex dolls. March 28. https://www.cps.gov.uk/cps/news/cps-takes-action-tackle-childlike-sex-dolls.

Danaher, John. 2017. Should we be thinking about robot sex? In *Robot sex: Social and ethical implications*, ed. John Danaher and Neil McArthur, 3–14. Cambridge: MIT Press.

David Hanson. 2021. Hanson robotics. https://www.hansonrobotics.com/david-hanson/.

Descartes, Rene. 1664. Treatise on man. Selection translated from *De l'homme et de la formation du foetus*, ed. Claude Clerselier, Paris, trans. P. R. Sloan. http://www.coretexts.org/wp-content/uploads/2010/08/DescartesTreatiseMnfin.pdf.

Devlin, Kate. 2018. *Turned On Science Sex and Robots*. London: Bloomsbury.

Diamonds, Ally Marie, and Wahine Toa Rising. 2021. Submission to South Australian Upper House Select Committee into decriminalisation of prostitution (Committee). https://www.parliament.sa.gov.au/api/sitecore/search/GetCommitteeFileDownload?id=378&fileId=1bc67c19-80d8-426f-bf1f-090a9bb7a109.

Di Nucci, Ezio. 2011. Sexual rights and disability. *Journal of Medical Ethics* 37 (3, March): 158–161.

Di Nucci, Ezio. 2017. Sexual rights, disability and sex robots. In *Sex Robots*, ed. John Danaher and Neil McArthur. Cambridge, MA: MIT Press.

Dworkin, Andrea. 1974. *Woman Hating*. NY: Plume.

Dworkin, Andrea. 1994. Prostitution and male supremacy. Speech at a symposium 'Prostitution: From Academia to Activism' held in 1992. *Michigan Journal of Gender and Law* 1 (1). https://repository.law.umich.edu/mjgl/vol1/iss1/1/?utm_source=repository.law.umich.edu%2Fmjgl%2Fvol1%2Fiss1%2F1&utm_medium=PDF&utm_campaign=PDFCoverPages.

Ekman, Kaijsa Ekis. 2013. *Being and being bought*, trans. Suzanne Martin Cheadle. North Melbourne: Spinifex.

European Parliament. 2017. Report with Recommendations to the commission on civil law rules on robotics. https://www.europarl.europa.eu/doceo/document/A-8-2017-0005_EN.html.

Floridi, Luciano. 2021. The European legislation on AI: A brief analysis of its philosophical approach. *Philosophy & Technology* 34: 215–222.

Fosch-Villaronga, Euard, and Adam Pooulsen. 2021. Sex robots in care. Setting the Stage for a Discussion on the Potential Use of Sex Robots for Persons with Disabilities,' *HRI '21 Companion: Companion of the 2021 ACM/IEEE International Conference on Human-Robot Interaction*, March 2021, 1–9. https://doi.org/10.1145/3434074.3446907.

Frye, Marilyn. 1983a. Oppression. In *The politics of reality*, 56–157. Crossing Press.

Frye, Marilyn. 1983b. To be and be seen. In *The politics of reality*, 56–157. Crossing Press.

Fung, Pascale. 2019. This is why AI has a gender problem. *Sustainable Development Impact Summit,* World Economic Forum, June 30. https://www.weforum.org/agenda/2019/06/this-is-why-ai-has-a-gender-problem/?fbclid=IwAR0rSI4_5fzYd4KOXAKbfvThgJBpw8fkOdEFLVtLaw8QByof7pY_i5tQTqI.

Garvey, Shunryu Colin. 2021. Unsavory medicine for technological civilization: Introducing 'Artificial Intelligence & its Discontents'. *Interdisciplinary Science Reviews* 46 (1–2, March): 1–18. https://doi.org/10.1080/03080188.2020.1840820.

Gordon, John-Stewart, and Ausrine Pasvenskiene. 2021. Human rights for robots? A literature review. *AI and Ethics* 1: 579–591. https://doi.org/10.1007/s43681-021-00050-7?utm_source=xmol&utm_medium=affiliate&utm_content=meta&utm_campaign=DDCN_1_GL01_metadata.

Graveris, Dainis. 2019. The best sex dolls for men in 2021: The definitive guide. September 30. https://sexualalpha.com/best-cheap-sex-dolls-for-men/.

Gritt, Emma. 2018. Sex robot creator Sergi Santos says people will marry his creations within the next 20 years–and claims he's worked out how to have a BABY with one. *The Sun*, October 5. https://www.thesun.co.uk/tech/4713921/sex-robot-creator-sergi-santos-says-people-will-marry-his-creations-within-the-next-20-years-and-claims-hes-worked-out-how-to-have-a-baby-with-one/#:~:text=Exclusive-,Sex%20robot%20creator%20Sergi%20Santos%20says%20people%20will%20marry%20his,have%20a%20BABY%20with%20one&text=SEX%20robot%20inventor%20Sergi%20Santos,society%20as%20we%20know%20it.

Haas, Benjamin. 2017. Chinese man 'marries' robot he built himself. *The Guardian*, April 4. https://www.theguardian.com/world/2017/apr/04/chinese-man-marries-robot-bobotuilt-himself.

Harmony. 2022. https://www.realdoll.com/product/harmony-x/.

Hauskeller, Michael. 2017. Automatic sweethearts for transhumanists. In *Robot Sex. Social and Ethical Implications*, ed. John Danaher and Neil McArthur. Cambridge, MA and London, England: MIT Press.

Hawthorne, Susan. 2020. *Vortex The Crisis of Patriarchy*. North Geelong: Spinifex Press.

Hill, Jess. 2019. *See What You Made Me Do: Power, Control and Domestic Abuse*. Black Inc, Carlton.

Hollandbeck, Amy. 2021. In a word: The dark magic of grammar. *The Saturday Evening Post*, March 4. https://www.saturdayeveningpost.com/2021/03/in-a-word-the-dark-magic-of-grammar/.

Huyssen, Andreas. 1982. The vamp and the machine: Technology and sexuality in Fritz Lang's metropolis. *New German Critique* (Autumn 1981–Winter 1982), 24/25, Special Double Issue on New German Cinema 221–237.

Illes, Judy, and Farhad R. Udwadia. 2019. Sex robots increase the potential for gender-based violence. *The Conversation*. https://theconversation.com/sex-robots-increase-the-potential-for-gender-based-violence-12236[illegible].

Jasanoff, Shelia. 2015. Future imperfect: Science, technology, and the imaginations of modernity. In *Dreamscapes of modernity: Sociotechnical imaginaries and the fabrication of power*, ed. S. Jasanoff and S.-H. Kim. University Press Scholarship Online.

Jecker, Nancy S. 2021. Nothing to be ashamed of: sex robots for older adults with disabilities. *Journal of Medical Ethics* 47 (1). https://jme.bmj.com/content/47/1/26.

Jeffreys, Sheila. 2008. Disability and the male sex right. *Women's Studies International Forum* 31: 327–335.

Jeffreys, Sheila. 2014. *Gender Hurts a Feminist Analysis of the Politics of Transgenderism*. Routledge.

Jordon, John M. 2019. The Czech play that gave us the word 'robot', *Medium*, *a*dapted from John Jordon's book *Robots*, part of the MIT Press Essential Knowledge series, MIT Press, July 29. https://medium.com/@mitpress/the-czech-play-that-gave-us-the-word-robot-86d4fdca003.

Kang, Minsoo. 2016. The mechanical daughter of Rene Descartes: The origin and history of an intellectual fable. *Modern Intellectual History* 14 (3): 633–660.

Keynote. Davecat. 2021. Speakers, 6th International Congress on Love & Sex With Robots 2021. https://www.lovewithrobots.com/speakers.

Kleeman, Jenny. 2017. 'The race to build the world's first sex robot. *The Guardian*, April 7. https://www.theguardian.com/technology/2017/apr/27/race-to-build-world-first-sex-robot.

Knox, Patrick. 2018. All dolled up. Meet Japanese porn star Asa Akira who inspired the world's most expensive sex robot EVER–and she'd even romp with it herself. *The Sun*, October 5. https://www.thesun.co.uk/news/4144351/japanese-porn-star-asa-akira-sex-robot/.

Kubes, Tanja. 2019. New materialist perspectives on sex robots. A feminist dystopia/utopia? *Social Science* 8 (8): 224.

Lennon, Kathleen. 2015. *Imagination and the Imaginary*. London: Routledge.

Lerner, Gerda. 1986. *The Creation of Patriarchy*. NY: Oxford University Press & Oxford.

Levy,. 2007. *Love and Sex with Robots: The Evolution of Human Robot Relationships*. NY: HarperCollins.

Liautaud, Susan. 2019. David Hanson: On Sophia means Wisdom... Dreams of what machines may be. *The Ethics Incubator*, June 25. https://www.youtube.com/watch?v=MFu0ksO_Wic.

Lockett, Jon. 2017. US military will have more combat robots than human soldiers by 2025. *New York Post*, June 15. https://nypost.com/2017/06/15/us-military-will-have-more-combat-robots-than-human-soldiers-by-2025/.

Mackinnon, Catharine A. 1987. Desire and power. *Feminism Unmodified. Discourses on Life and Law*. Cambridge, MA & London: England Harvard University Press.

Maloney, Alison. 2018. Sex robot creator claims "five minute romps" with his lifelike android have "improved his marriage. *The Sun*, October 2. https://www.thesun.co.uk/news/4592234/sex-robot-creator-claims-five-minute-orgasms-with-his-lifelike-android-has-improved-his-marriage/.

Maloney, Alison. 2022. 'Sex robot creator claims "five minute romps" with his lifelike android have "improved his marriage".' *The Sun*, October 2. https://www.thesun.co.uk/news/4592234/sex-robot-creator-claims-five-minute-orgasms-with-his-lifelike-android-has-improved-his-marriage/.

Maslow, A.H. 1943. A theory of motivation. *Psychological Review* 50: 370–396.

Merchant, Carolyn. 1980. *The Death of Nature. Women, Ecology and the Scientific Revolution*. NY: Harper & Row.

Miailhe, Nicolas, and Cyrus Hodes. 2017. The third age of artificial intelligence. *Field Actions Science Reports: Artificial Intelligence and Robotics in the City* 17: 6–11. https://journals.openedition.org/factsreports/4383.

Millet, Kate. 1970. *Sexual Politics*. Urbana & Chicago: University of Illinois Press.

Misa, Thomas J. 2003. The Compelling Tangle of Modernity and Technology. In *Modernity and Technology*, 1–3. Cambridge, MA: MIT Press.

Moran, Rachel. 2013. *Paid For. My Journey Through Prostitution*. Gill & Macmillan.

Morgan, Robin. 2014. *Going Too Far, The Personal Chronicle of a Feminist*. NY: Open Road Integrated Media.

Negri, Sergio M. C. Avila. 2021. Robot as legal person: Electronic personhood in robotics and artificial intelligence. *Frontiers in Robotics and AI. Ethics in Robotics and Artificial Intelligence*, December 23. https://doi.org/10.3389/frobt.2021.789327/full.

O'Connor, Patrick. 2018. On sex robots and personhood with Kathleen Richardson. *Thales' Well*, podcast, October 4. https://podcastaddict.com/episode/114752143.

Open Society Institute. 2006. *Harm Reduction, Health and Human Rights, and Sex Work*. NY: Sexual Health and Human Rights Project. https://www.opensocietyfoundations.org/uploads/1438121e-9fbf-4875-9e9c-b6e50221652d/approaches_20060601.pdf.

Owsianik, Jenna 2022. *State of the Sexbot Market: The world's best sex robot and AI sex doll companies*. Future of Sex. https://futureofsex.net/robots/state-of-the-sexbot-market-the-worlds-best-sex-robot-and-ai-love-doll-companies/

Papadaki, Evangelia (Lina). 2019. Feminist perspectives on objectification. *Stanford Encyclopedia of Philosophy*. https://plato.stanford.edu/entries/feminism-objectification/.

Pateman, C. 1988. *The Sexual Contract*. Cambridge: Polity Press.

Pateman, Carole, and Charles W. Mills. 2007. *Contract and Dominationi*. UK & Malden MA, USA: Polity Press Cambridge.

Pride Flag Identification Guide. 2021. https://www.prideflags.org.

Qvist, Anastasia. 2020. Possession of childlike sex dolls. *Armstrong Legal*. https://www.armstronglegal.com.au/criminal-law/nsw/offences/sexual/possession-of-child-like-sex-dolls/.

Rajnerowicz, Kazimierz. 2021. Will AI take your job? Fear of AI and AI trends for 2021. *TIDIO*. https://www.tidio.com/blog/ai-trends/.

Real Sex Dolls. 2022. https://real-sex-dolls.com/.

Riccio, Thomas. 2021. Sophia robot. An emergent ethnography. *The Drama Review* 65: 3. https://www.cambridge.org/core/journals/the-drama-review/article/sophia-robot/724ECA22CFC406398D70DE0443B2FF0F.

Rich, Adrienne. 1980. Compulsory heterosexuality and lesbian existence. *Signs* 5 (4) (Summer 1980); Women: Sex and Sexuality (Summer): 631–660.

Rich, Adrienne. 1995. *Of Woman Born. Motherhood as Experience and Institution*. NY & London: W. W. Norton.

Richardson, Kathleen. 2016. Sex robot matters. *IEEE Technology and Society Magazine*, June. https://ieeexplore.ieee.org/document/7484884.

Richardson, Kathleen. 2021. Sex, robots and 'artificial intimacy': A threat or the next best thing? Presentation at *Is it Time to have Sex with Robots?* Centre for Humans and Technology, Eindhoven University of Technology, May 31. https://www.cursor.tue.nl/en/news/2021/mei/week-2/is-it-time-to-have-sex-with-robots/?fbclid=IwAR1yn_EReM4IJz-vnX18m_JDKbYbtaQ4USgkmk2BOZP3PtVzAUK7HnhrHQ0.

Robosexuality. n.d. MOGAI Encyclopedia. https://rainbowpedia.fandom.com/wiki/Robosexuality#:~:text=A%20robosexual%20(or%20technosexual)%20is,be%20of%20any%20sexual%20orientation.

Sohrabi-Shiraz, Ariane. 2018. Sex robot maker Sergi Santos: Who is he and what does Synthea Amatus do? *Daily Star*, January 9. https://www.dailystar.co.uk/news/latest-news/sex-robot-sergi-santos-who-16837966.

Sophia. 2021. Hanson robotics. https://www.hansonrobotics.com/sophia/.

Srinivasan, Amia. 2021. *The Right To Sex*. London & Dublin: Bloomsbury.

Stevenson, Daniel C. 1994–2000. Ovid, *Metamorphoses, Book the Tenth*, Translated by Sir Samuel Garth, John Dryden, et al. *The Internet Classics Archive*. http://classics.mit.edu/Ovid/metam.mb.txt.

Switzky, Lawrence. 2020. ELIZA effects: Pygmalion and the early development of artificial intelligence. *Shaw: The Annual of Bernard Shaw Studies* 40 (1): 50–68.

Taylor, Lin. 2017. Sex robots: Perverted or practical in fight against sex trafficking? *Reuters*, July 20. https://www.reuters.com/article/us-science-robots-sextrafficking-idUSKBN1A50HD.

Touching Base Inc 2022. https://www.touchingbase.org/.

Turkle, Sherry. 2021. *The Empathy Diaries: A Memoir*. London: Penguin Press.

UN Office on Drugs and Crime. 2019. Global Study on Homicide. https://www.unodc.org/unodc/en/data-and-analysis/global-study-on-homicide.html.

UN Women. 2021. Facts and figures: Ending violence against women. March. https://www.unwomen.org/en/what-we-do/ending-violence-against-women/facts-and-figures.

UNDP. 2021. UNDP in Asia and the Pacific Appoints World's First Non-Human Innovation Champion. *United Nations Development Program*. https://www.asia-pacific.undp.org/content/rbap/en/home/presscenter/pressreleases/2017/11/22/rbfsingapore.html.

Urbi Jaden, and MacKenzie Sigalos. (2018). The complicated truth about Sophia the robot—an almost human robot or a PR stunt. *CNBC*, June 5. https://www.cnbc.com/2018/06/05/hanson-robotics-sophia-the-robot-pr-stunt-artificial-intelligence.html.

Vaswani, Karishma. 2021. Alibaba accepts record China fine and vows to change. *BBC News*, April 12. https://www.bbc.com/news/business-56713508.

Vice Media. 2018. The first sex doll brothel in Europe. YouTube. https://www.youtube.com/watch?v=xcBJ-vA5AoE.

Willis, Ellen. 2012. 'Lust Horizons: Is the Women's Movement Pro-Sex?' In E. Willis (Ed.). *No more nice girls. Countercultural Essays*. Minneapolis & London: University of Minnesota Press.

Winner, Langdon. 1980. Do artifacts have politics? *Daedalus* 109 (1): 121–136.

Wiseman, Eva. 2015. Sex, love and robots: is this the end of intimacy? *The Guardian*, December 13. https://www.theguardian.com/technology/2015/dec/13/sex-love-and-robots-the-end-of-intimacy.

Women's Declaration International. 2022. Declaration on Women's Sex-Based Rights: Full Text. https://www.womensdeclaration.com/en/declaration-womens-sex-based-rights-full-text/#article-1.

Zarkodakis, George. 2013. Love machines. *aeon*, March 26. https://aeon.co/essays/the-lure-of-ai-is-erotic-as-much-as-it-is-rational.

CHAPTER 7

Paedophilia, Child Sex Abuse Dolls and the Male Sex Right: Challenging Justifications for men's Sexual Access to Children and Child Sexual Abuse Material

Caitlin Roper

The BBC Three documentary *The Future of Sex: Sex Robots and Us* follows presenter James Young as he interviews manufacturers of lifelike sex dolls and robots around the world (BBC 2018). In a Tokyo sex doll factory, with headless dolls at varying stages of development strewn all around, one doll in particular catches Young's attention. Unlike the other dolls modelled on adult women, this one is small.

Visibly uncomfortable and blinking back tears, Young asks, "What is this doll?" Hiro Okawa, the manufacturer, responds, "We have to leave

C. Roper (✉)
Collective Shout, Melbourne, Australia
e-mail: caitlin@collectiveshout.org

© The Author(s), under exclusive license to Springer Nature Switzerland AG 2022

K. Richardson and C. Odlind (eds.), *Man-Made Women*, Social and Cultural Studies of Robots and AI,
https://doi.org/10.1007/978-3-031-19381-1_7

it to the customer's imagination for the actual age setting." Apparently sensing Young's distress, he adds, "Of course, I understand what you are trying to say. But there might be some kind of sentiment to this petite, kid-like size." With a shaking hand, Young wipes his eyes. Once outside, he describes the experience as "horrific"—he "had to get out of there."

This doll was not a one-off. Lifelike, anatomically correct silicone dolls modelled on the bodies of children, complete with penetrable orifices and marketed for men's sexual use—are already on the market. A number of jurisdictions have criminalised child sex abuse dolls on the basis that they encourage the sexual abuse of children but advocates for the products argue they could be made available for men's paedophilic use (Behrendt 2017, Cantor 2017, 2018; Moen and Sterri 2018; Prostasia Foundation 2019a).

This chapter examines the ways in which paedophilic men advocate their access to child sex abuse dolls by building on their historical justifications and warns against further development of AI enhancement into childlike sexbots. While attitudes towards child sexual abuse have evolved, the underlying pitch by their advocates remains the same. Paedophilic men have strategically framed their sexual use of both children and child sex abuse dolls as being in children's best interests, rather than their own. They have aligned their use and abuse of children with sexual liberation, and derided feminist objections as sexual repression (Angelides 2004; Thorstad 1979 as cited in Rush 1980). Narratives of victimhood are often drawn upon by men who sexually abuse children and their supporters painting opposition as a conservative moral panic and framing laws preventing men's sexual access to children as the unwelcome regulation of sexual behaviour and an attack on their sexual rights (Califia 1994; Rubin 1984).

I argue there is no evidence for the claim that men's access to child sex abuse dolls can prevent their abuse of living children. On the contrary, child sex abuse dolls could *increase* the risk of child sexual abuse. Men who are found in possession of a child sex abuse doll are typically also found with child sexual abuse material, the filmed abuse of children, or to be sexually offending against children in additional ways—even incorporating children into their child sex abuse doll use. I call for a multi-pronged approach to this technology, including uniform government legislation to criminalise child sex abuse dolls, and for online marketplaces to block the sale of these products on their platforms.

A History of Defending Men's Sexual Access to Children

In order to contextualise current arguments in support of child sex abuse dolls, I will first provide background history which is vitally important to understanding present-day justifications for the products. From past decades to the present, men have utilised narratives of child sexuality to justify their sexual use of children. In the 1970s, it was widely accepted in the field of psychiatry that children were sexual beings. Freud, the founder of psychoanalysis, believed that sexuality was present in all humans from infancy. "Father of the Sexual Revolution" Kinsey claimed to have documented sexual responses in infants under six months old—data which could only have been obtained through their sexual abuse (Reisman and McAlister 2018). But many researchers did not regard sexual abuse as inherently harmful to children, instead attributing any harm to societal stigma (Ramey 1979).

Some researchers argued that children were enthusiastic participants, desiring and enjoying sexual abuse, even casting them in the role of instigator or seductress (Bender and Blau 1937; Henderson 1975; James 1970). Men's sexual access to children was conveniently framed as an exercise in children's autonomy, i.e. an expression of their sexuality. Laws which prevented men's sexual use of children, such as age of consent laws, were condemned for repressing children's sexuality. For example, sexologist Edwin J. Haeberle called for the abolition of laws against incest on the basis that children were being denied "their right to sexual satisfaction" (Herman 1981: 25). The concept of children's 'right' to sexual expression was useful for paedophiles and their supporters, as it allowed them to shift the focus from men's sexual predation to children's apparent 'right' to engage in sexual interactions with them. As radical feminist scholar Sheila Jeffreys pointed out, "The children, of course, were not clamouring for sexual liberation – only the men who wanted to use them" (Jeffreys 2021 as cited in Clásicas y Modernas 2021).

Feminists who objected to men's sexual abuse of children were accused of repressing children's sexuality. Second-wave feminists exposed the phenomenon of incest and child sexual abuse, which they situated as a male abuse of power (Rush 1980). Angelides (2004) described how feminists had debunked the myth of 'stranger danger,' highlighting the prevalence of fathers and male relatives as perpetrators of abuse and citing the unequal power relations between adults and children. However, while acknowledging the significant work of feminists in addressing men's sexual abuse of children, Angelides accused them of disempowering children by contributing to the repression of their sexuality:

> The discourse of child sexual abuse has expanded at the expense of a discourse of child sexuality...The feminist use of power has functioned to evade, silence, erase, and repress a signifier of child sexuality. (p. 142)

Indeed, feminists' "utter disregard" for child sexuality was said to be "as damaging to a child's social and psychological well-being as the discounting of the reality of sexual abuse" (p. 158). In contrast to feminist objections characterised as sexual repression, men's sexual abuse of children has been portrayed as sexual liberation. Queer theorists and paedophile apologists have drawn on narratives of victimhood, framing paedophiles as an oppressed, stigmatised minority who are subjected to persecution, hatred and civil rights violations. Campaigning against laws criminalising paedophilia, founder of NAMBLA (North American Man/Boy Love Association) David Thorstad wrote,

> We are engaged in a war between the forces of sexual liberation on the one hand and the forces of sexual repression on the other. Man/Boy love and cross generational sex have become the cutting edge of that war. Repeal all age of consent laws!!! Freedom of sexual expression for all!!! (Thorstad 1979 as cited in Rush 1980: 188–189)

Despite paedophiles being almost exclusively adult men, and their victims children, paedophile defenders have inverted understandings of power structures to position *paedophiles* as the vulnerable group. The language used to describe paedophilia functions to obscure the reality of the power disparity between men and children, and to paint men's sexual abuse of children as a relationship. Previously, men who sexually abused children were euphemistically referred to as "boy lovers," "men who love underage youth" and "those whose eroticism crosses generational boundaries" (Rubin 1984: 145, 147, 151). Men's sexual abuse of children was white-washed as "cross-generational encounters" (p. 153) or "intergenerational sex" (Angelides 2004: 142), certainly more palatable than 'men who rape and sexually abuse children.'

Paedophiles and their supporters have portrayed objections to paedophilia, and laws criminalising it, as an attack on "erotic communities" (Rubin 1984: 145). They have argued that age of consent laws which prevented men's sexual access to children, and laws criminalising child pornography, represented the regulation of sexual behaviour, violation of civil liberties and the further victimisation of paedophiles. In *Thinking Sex* (1984), one of the foundational texts of queer theory, Gayle Rubin characterised paedophiles as the "victims of a savage and undeserved witch hunt" (p. 147), therefore casting men who sexually abuse children, and those who desire to do so, as the victims—not the children they rape and abuse, who are made invisible.

While male abusers were positioned as the legitimate victims of a conservative moral panic, the children they raped were relegated to the status of "invented" victims:

> When the furore has passed, some *innocent erotic group has been decimated*, and the state has extended its power into new areas of erotic behaviour. The system of sexual stratification provides *easy victims who lack the power to defend themselves*, and a pre-existing apparatus for controlling their movements and curtailing their freedoms. The stigma against sexual dissidents renders them morally *defenceless* ... Moral panics rarely alleviate any real problem, because they are aimed at chimeras and signifiers. They draw on the pre-existing discursive structure *which invents victims in order to justify treating 'vices' as crimes.* (Rubin 1984: 163, my emphasis added)

Furthermore, paedophile apologists argued men's sexual use of children was not inherently abusive or harmful, and that children are not necessarily victims, but are often willing participants:

> Any child old enough to decide whether or not she or he wants to eat spinach, play with trucks, or wear shoes is old enough to decide whether or not she or he wants to run around naked in the sun, masturbate, sit in somebody's lap, or engage in sexual activity. (Califia 1994)

Paedophile advocates argued the criminalisation of child pornography (now known as child sexual abuse material) was primarily an attack on the sexual rights of paedophiles. In the late 1970s in the US, a federal law was introduced which prohibited any depiction of children nude or engaged in sexual activity (Califia 1994). In 1990, a Supreme Court ruling criminalised the possession and viewing of child pornography (Califia 1994

[1991]). Defenders of paedophilia argued the laws against child pornography were unreasonable, preventing "even the private possession" of pornography involving children (Rubin 1984: 146). But, as Rubin noted critically, apart from NAMBLA and the American Civil Liberties Union, "no one raised a peep of protest" (p. 146).

Writer Pat (now Patrick) Califia lamented the "panic" over the "bogeyman" of child pornography or "kiddy-porn" (1994 [1991]). Califia claimed its prevalence had been exaggerated and that most of the content was not necessarily abusive, being mainly comprised of depictions of naked children and young people engaged in sexual activity with each other. Only a "very tiny proportion" showed adults "engaged in sexual contact with minors" and it would be "a mistake to characterise all child porn as 'a record of abuse'":

> Sometimes it was a record of children's exhibitionism and free erotic play with one another. Sometimes it was a record of adolescent vanity, pride, and budding sexuality. Sometimes it preserved a moment of exceptional trust and pleasure *between partners whose ages would normally have kept them apart.* (Califia 1994 [1991], my emphasis added).

The creation and distribution of child pornography was downplayed and dismissed as "boy-lovers" merely taking (pornographic) photos of their "youthful partners," images which they might share with their friends—and "doesn't everyone take pictures of loved ones?" (Califia 1994).[1]

In *Thinking Sex* (1984), Rubin recounted the case of photographer Jacqueline Livingston, who was fired and under threat of prosecution for taking and exhibiting photographs of her seven-year-old son masturbating. This was shared as evidence of the frightening climate created by the "child porn panic," with Rubin concluding, "It is easy to see someone like Livingston as a victim of the child porn wars. It is harder for most people to sympathise with actual boy-lovers" (p. 147). Again, paedophiles and their supporters have portrayed the adults (usually men) as stigmatised victims who risked punishment for producing child pornography—not the children used in its production.

[1] Pat Califia has since updated her position—see *Public Sex: The Culture of Radical Sex* (2000).

Contemporary Defences of Men's Access to Child Sexual Abuse Material

Attitudes towards child sexual abuse have evolved considerably since then. It is now widely understood that children cannot legally consent to sex acts with adults (World Population Review 2022) that sexual contact between adults and children constitutes sexual abuse (RAINN 2022) and that sexual abuse is demonstrably harmful to children (WHO 2003). The arguments previously made by queer theorists and paedophiles, regarding children's supposed rights to engage in sexual 'relationships' with adult men, would be unlikely to achieve mainstream acceptance today. It is therefore now much more difficult for paedophiles and queer theorists to claim men's sexual use of children is an expression of children's sexual autonomy, or that laws against child sexual abuse are an attack on paedophiles' civil liberties. As a result, their tactics have changed.

Now a therapeutic model of paedophilia as a sexual or psychiatric disorder bolsters persistent narratives of paedophiles as victims. There is a concerted effort to destigmatise paedophilia and disentangle it from child sexual abuse (Berlatsky 2021; Cantor 2019a; Earp 2017; Harper et al. 2021; Walker 2021). Some academics understand paedophilia as a legitimate and unchangeable sexual attraction or orientation, outside the control of the individual—a condition that a person (usually a male) is afflicted with (Cantor 2019b; Moen and Sterri 2018; Seto 2012). For example, some academics and advocates paint paedophiles as victims of their circumstances who are deserving of sympathy, compassion and understanding, lamenting the "unfortunate situation in which they find themselves" (Moen and Sterri 2018: 377).

Paedophiles have rebranded as Minor-Attracted Persons (MAPs). Some describe themselves as non-offending, non-contact or even "virtuous" claiming they do not sexually abuse children (Virped 2012). Paedophiles still profess to be an oppressed minority (see Elchuk et al. 2021) but now on the basis that—through the therapeutic model—paedophilia is understood as a condition, not a choice (Moen and Sterri 2018) rather than the feminist conception of paedophilia and child abuse as a male abuse of power (Rush 1980).

A number of academics and advocates (predominantly male) argue, firstly, that since paedophiles cannot ethically act on their sexual desires to rape and abuse children, they could be allowed access to 'sexual outlets' that allow them to simulate it (Behrendt 2017; Cantor 2019 in Nicol

2019; Harper and Lievesley 2022; Moen and Sterri 2018; Prostasia Foundation 2019a; Prostasia n.d.). These include so-called virtual or artificial child sexual abuse material (computer-generated or artistic depictions of non-existing children produced without the use of children) as well as child sex abuse dolls. Secondly, some paedophile advocates and academics go even further, claiming that providing paedophiles with childlike dolls on which to enact their fantasies of raping children could prevent their abuse of children (Behrendt 2017; Cantor 2019 in Nicol 2019; Levy 2015 in Wiseman 2015; Harper and Lievesley 2022; Moen and Sterri 2018; Prostasia Foundation 2019a, Prostasia n.d.). For example, David Levy, author of *Love and Sex with Robots* (2007) told *The Guardian*, "It would be better for paedophiles to use robots as their sexual outlets than to use human children" (Levy 2015 as quoted in Wiseman 2015). Others have suggested that providing paedophiles with child sex abuse dolls could not only prevent their sexual offending against children, but that failing to do so could put children at risk. In his 2017 paper proposing "child sexbots" be made available to paedophiles for "therapeutic" use, philosopher Marc Behrendt described the potential cost of not providing these products to paedophiles:

> If the end justifies the means … any kind of therapeutic option should be considered, analysed, scrutinised, weighed and even tried … can civil society still afford to take the risk of seeing other unfortunate victims, unwittingly hooked to the appalling web of sex offenders? Other souls traumatised for life? Other murdered children? (Behrendt 2017: 97)

Likewise, Canadian psychologist and neuroscientist James Cantor asks, "If we surveyed victims of childhood sexual abuse, I wonder how many would say they WISHED their abusers had a sex doll who might have taken their place?" (Cantor 2019 as quoted in Nicol 2019).

Finally, where paedophilic men previously portrayed their sexual access to children as being in children's best interests, they now portray their access to child sex abuse dolls as preventative (Behrendt 2017; Harper and Lievesley 2022; Malcolm 2021d; Moen and Sterri 2018). Defenders of these technologies do not use the rhetoric of men's sexual rights, which is unlikely to be a winning strategy. Instead, they argue these products could protect children from sexual abuse, and purport to be motivated by a concern for their safety and well-being. Framing child sex abuse dolls as potentially preventing child sexual abuse is an approach much more likely to garner support, even in the absence of any empirical evidence to support their claims.

Defenders of virtual child sexual abuse material and child sex abuse dolls accuse feminists who object of being motivated by a "moral purity agenda" rather than out of any meaningful concern for children (Malcolm 2021b). For example, Jeremy Malcolm, executive director of Prostasia Foundation, which bills itself as a child protection organisation but campaigns against laws criminalising child sex abuse dolls and sexualised depictions of children in drawing and cartoons, objects to sex offender registries and funds research into "fantasy sexual outlets" for paedophiles (Slatz 2021). Unlike other child protection organisations, Prostasia promotes ageplay, BDSM and kink, and "centers its work around safeguarding the civil liberties of adults" (Férdeline 2020). Malcolm accuses feminists who campaign against child sex abuse dolls—referred to by Prostasia Foundation (2021) as "purity campaigners"—of having a secret agenda to police sexual practices they disapprove of *("Everything but straight, cis, missionary sex is deviant in your book")* and an unhealthy interest in other people's sex lives (*"your obsession with how other people masturbate is honestly invasive and creepy"*) (Malcolm 2021c). Malcolm portrays feminist objections as purely ideological and unscientific—*"moral purity is more important to you than possibly preventing abuse"* (Malcolm 2021d)—in contrast to those, like him, who support the development of child sex abuse dolls for men's sexual use as rational, scientific and on the side of sexual liberation. Unsurprisingly, they are almost entirely, men.

Critical Responses to Common Arguments in Support of Child Sex Abuse Dolls

"It's just a sex toy like any other"

One of the arguments made in support of child sex abuse dolls is that they are sex toys like any other. Prostasia Foundation characterises men's use of child sex abuse dolls as "the personal and private use of sex toys" and claims that laws prohibiting them are therefore "unconstitutional, misguided and immoral" (Prostasia Foundation 2019a) James Cantor tweeted, "I'm okay with latex sex toys and I don't care what they look like" (Cantor 2018) and "Australia protecting society from crimes against latex. Thought crimes against latex" (Cantor 2017).

If child sex abuse dolls are conceptualised as merely sex toys like any other, it could theoretically be argued they do not raise any unique ethical issues. But child sex abuse dolls, like sex dolls and robots more generally, are distinct from traditional sex toys on the basis of their embodiment in human form—most often, the female form. They are not simply latex masturbation aids, they are lifelike, anatomically correct representations of girls with penetrable orifices and marketed for men's sexual use. The childlike appearance is not incidental, it is the very point. Child sex abuse dolls are intentionally designed to look and feel like a real child, to facilitate men's fantasy experience of raping a little girl. If the products were truly nothing more than a latex sex toy, there would be no need for them to be shaped like little girls in the first place.

The marketing of child sex abuse dolls is revealing. Online photo galleries depict dolls designed to resemble pre-pubescent girls styled with pigtails, dressed in clothing emblazoned with popular children's characters and holding soft toys. Often the camera focuses on the chest or crotch regions, with dolls in varying states of undress, or naked. Others are costumed for seduction in lingerie complete with fishnet stockings and whips, or shown laying supine, with heart-shaped patterns of whipped cream and strawberries concealing the genitals.

Some dolls are clearly marketed in such a way as to eroticise predation on children (Roper 2020a). The most chilling images are those that appear to replicate scenarios of child abuse. A photograph in one popular doll manufacturer photo gallery portrays a petite doll that appears as a girl aged about six or seven, lying face down on a bed with underpants pulled down. Child sex abuse dolls come with a range of different facial expressions. Some appear to be smiling, others appear to be crying or in pain. Shin Takagi, owner of child sex abuse doll manufacturer Trottla, explains this is done "to fulfil a variety of client needs" (Morin 2016). The ability to simulate child sexual abuse is precisely the drawcard for users. Indeed, this is why some academics pitch child sex abuse dolls as an alternative to sexually abusing a child, because they believe the dolls could function as surrogates for children. Equating men's use of child sex abuse dolls modelled on the bodies of little girls with the use of a sex toy is a deliberate attempt to decontextualise men's simulation of child sexual abuse from a wider culture in which men sexually abuse girls at an alarming rate—with roughly 20% of girls estimated to be subjected to sexual abuse in Australia (Gilmore 2017).

"It's a victimless crime – you can't rape a robot"

Defenders of men's access to child sex abuse dolls and robots argue that given they are objects, not moral agents, they cannot be victims of rape or child sexual abuse and conclude there is no harm: you can't rape a robot (Malcolm 2021a, Prostasia Foundation 2019b). But feminist objections are not focused on harm to dolls, robots, or inanimate objects—they are concerned with harm to children.

Advocates argue that men's sexual use of child sex abuse dolls is a victimless crime in the same way they have argued men's consumption of virtual child sexual abuse material is a victimless crime, because no child is being abused in its production (Malcolm 2021e). But legitimising children as appropriate objects of male sexual desire and gratification puts them at risk, normalising men's sexual use and abuse of children and encouraging offenders. According to the United Nations Report of the Special Rapporteur on the sale and sexual exploitation of children, including child prostitution, child pornography and other child sexual abuse material 2020, sexualised depictions of "non-existing" children serve to normalise men's sexual abuse of children:

> The increased accessibility and availability of child sexual abuse material online appears to normalise this crime and may encourage potential offenders and increase the severity of abuse. This includes new phenomena, such as drawings and virtual representations of non-existing children in a sexualised manner, widely available on the Internet.
>
> The increasing social acceptance of early sexualisation is exacerbated by the widespread dissemination of child sexual abuse material on the Internet and the production of highly realistic representations of non-existing children. This objectification of children comforts offenders in their actions. (UN Special Rapporteur 2020: 11–12)

Similarly, feminist sociologist Gail Dines argued that even where actual children are not used in its production, pornographic content portraying children undermines important social norms that define children as off limits for men's sexual use. Writing on 'Pseudo Child Pornography' (also known as 'Teen' or 'Barely Legal' porn) where adult women are depicted as children and styled with braces and pigtails, Dines described how the eroticisation of children could influence male consumers' attitudes towards children and put them at risk:

> Once they click on these sites, users are bombarded—through images and words—with an internally consistent ideology that legitimises, condones, and celebrates a sexual desire for children. The norms and values that circulate in society and define adult-child sex as deviant and abusive are wholly absent in PCP [Pseudo Child Pornography] and in their place a cornucopia of sites that deliver the message (to the viewer's brain, via the penis) that sex with children is hot fun for all ... There is a wealth of research within media studies that shows people construct their notions of reality from the media they consume, and the more consistent and coherent the message, the more people believe it to be true ... This does not mean that all men who masturbate to PCP will rape a child, or even be sexually attracted to a child. What it does mean, however, is that on a cultural level, when we sexualise the female child, we chip away at the norms that define children as off limits to male sexual use. (Dines 2011: 7–8)

"They could prevent child sexual abuse"

Perhaps the most common claim in support of child sex abuse dolls is that they could prevent child sexual abuse (Behrendt 2017; Harper and Lievesley 2022; Malcolm 2021d; Moen and Sterri 2018). However, a 2019 report from the Australian Institute of Criminology found not only that there is no evidence child sex abuse dolls could prevent sexual abuse, but that they could *increase* the risk of child sexual abuse. Authors Rick Brown and Jane Shelling warned that child sex abuse doll use could lead to an escalation in sexual offences against children, desensitise users to the harm of sexual abuse (as dolls do not offer any emotional feedback), be used to groom children for abuse, and "promote a continuum of behaviour that results in contact offending, by bridging the gap between fantasy and reality" (Brown and Shelling 2019: 4).

Men's sexual preferences—even their desires to rape children—are elevated to the status of 'needs' that must be met. Underpinning calls to make child sex abuse dolls available to paedophilic men is the acceptance of men's sexual preferences as absolute; the belief they have a right to sex in any way they want it, and that anything less is deprivation. The academics and advocates who support the development of child sex abuse dolls are so deeply committed to their belief in the male sex right that their answer to men who want to sexually abuse children is not to challenge their sexual appetites, not to encourage the pursuit of healthier practices or relationships that do not involve children, but to endorse the manufacture of replica children with penetrable orifices so they can enact their rape fantasies unimpeded.

Child sex abuse doll advocates fail to consider the wider cultural context in which men sexually offend against girls—a system of institutionalised male dominance and a culture that eroticises girls. Girls are routinely presented as sexually available and appealing. Many women report first experiencing sexual attention from adult men as girls. 'Teen porn' is consistently one of Pornhub's most popular search terms (Pegg 2016). 'Barely Legal' pornography depicting teens with pigtails, flat chests and braces is widely available (Tankard Reist 2008). School girls are fetishised, and sexy school girl costumes are sold in mainstream retailers (Roper 2015). Major bookstores and online marketplaces like Amazon (Hall 2019) and Bookworld (Collective Shout 2013) have been exposed selling 'erotic' eBooks that feature incest and child sexual abuse. Sexualised images of underage and even pre-pubescent girls can easily be found on social media apps like Instagram, typically met with predatory sexual comments from adult men (Kennedy 2019).

Men's demand for child sex abuse dolls modelled on the bodies of pre-pubescent girls cannot be separated from a wider culture in which girls are sexualised and treated as objects of men's sexual entertainment. Rather than finding more avenues to accommodate men's sexual desires for girls, a better approach is to address the roots of these desires, to challenge the cultural factors at play that encourage and legitimise men's sexual entitlement and predation on girls.

Men Who Own Child Sex Abuse Dolls Still Sexually Offend Against Children

The approach of child sex abuse dolls as preventative therapy does not stand up to evidence. The reality of men's use of child sex abuse dolls is not as simple as a doll or a child. In the UK, seizures of child sex abuse dolls have led investigators to identify previously unknown sexual offenders against children (Rawlinson 2017). Men who are found in possession of a child sex abuse doll are typically also found in possession of images and video of child sexual abuse material, featuring the abuse, rape and torture of children (Roper 2020b).

South Australian man Shane Andrew Lunnay, who live-streamed himself having sex with a child sex abuse doll, was found with drawings and animated content depicting the sexual abuse of children aged between 3 and 12, as well as images of real children being sexually abused (Bermingham 2020). In England, former soldier Richard Turner was found with a child sex abuse doll as well as 100,000 images of child sexual abuse material (203 of these being the most serious kind) and pleaded guilty to ten charges of sexually offending against children (Hetherington 2019). In the US, James Randall Adams was found with 62, 381 images and 738 videos on his phone which were mainly child sexual abuse material, as well as an adult female-bodied mannequin and a bound child mannequin that both appeared to be used as sex dolls (Reavy 2021b). An unnamed Utah man was found with eight child sex abuse dolls, two of them dressed in the gown from Disney's Frozen, after a social media company reported him to police for uploading and distributing child sexual abuse material (Reavy 2021a). A police raid of the home of New South Wales man Stephen Capsis turned up a child sex abuse doll and a "trove" of child sexual abuse material. Capsis had been caught at the beach filming intimate videos of young girls in bikinis and zooming in on their genitals using a covert device attached to his car (Lockley and Barr 2021). Owning a child sex abuse doll or consuming virtual child sexual abuse material did not prevent these men, nor any of the others in a growing number of reported cases like these, from sexually offending against children.

There is also evidence of male owners of child sex abuse dolls incorporating living children into their doll usage. Police charged Adelaide man Adam Neil Wonnocott after they caught him with hundreds of images of child sexual exploitation material, among which the faces of children known to him had been superimposed over the faces of abused children. He also owned five home-made child sex abuse dolls. One of these dolls, dressed in girls' clothing and with a sex toy built into it, had a laminated photo of a real child's face attached to its head (Mott 2021). Victorian primary school teacher James William Treasure filmed himself having intercourse with a child sex abuse doll in a school uniform with masks made from school photos of his students. Police also found videos of sex acts on which Treasure had digitally imposed his own face and that of one of his young students, as well as thousands of child abuse images and several covert recording devices (Hobday 2021).

Doll manufacturers, themselves, have created child sex abuse dolls in the likeness of actual children. In 2020, Child Rescue Coalition reported that a Florida woman discovered a child sex abuse doll on Amazon apparently modelled on her eight-year-old daughter. One image of the doll appeared to recreate that of her daughter in a photo previously shared to social media (Child Rescue Coalition 2020). A Collective Shout investigation of global online marketplace Etsy has exposed doll manufacturers who offer to customise child sex abuse dolls to look like specific children. One seller claimed a particular doll was modelled on a 14-year-old Instagram model (Hall 2020). Child sex abuse dolls do not just aid in the fantasy of abusing a generic, hypothetical child; they allow users to victimise a *specific* child without even having physical access to them. Far from being a 'victimless' crime, the development of child sex abuse dolls facilitates a new form of technologically mediated sexual abuse, one where men can simulate the rape, abuse and torture of specific children using dolls made in their likeness.

A number of major online platforms have been exposed for facilitating the sale of child sex abuse dolls and replica child body parts marketed for men's sexual use. Grassroots campaigning movement Collective Shout successfully lobbied both budget shopping app Wish (Liszewski 2018) and Chinese e-commerce platform Alibaba (Collective Shout 2020) to remove such products, and is engaged in an ongoing campaign against online marketplace Etsy (Roper 2022). In response to Collective Shout's campaigning efforts, Alibaba removed all child sex abuse dolls from sale, and shortly after, geo-blocked the sale of all sex dolls to Australia (Collective Shout 2021).

To combat the harms of child sex abuse dolls, a multi-pronged approach is necessary. Global online marketplaces which have hosted child sex abuse doll listings need to implement strong policies and practices that prevent the sale of these products on their platforms, permanently banning sellers in breach of terms and conditions and passing their details on to law enforcement. They need to suffer penalties if they do not. Governments too must act urgently to uphold the rights of children, among the most vulnerable members of society, through uniform legislation criminalising the manufacture, sale, importation and possession of child sex abuse dolls. There must be significant penalties to these offences to reflect cultural values that children are off limits for men's sexual use.

Conclusion

The child sex abuse doll market is still in its infancy. Nevertheless, children have already been harmed through their manufacture and use. Australian Border Force has reported a 653% increase in detections of child sex abuse dolls over the past two years (Chung 2021). Experts have predicted that child sex abuse robots will be next, and already may be in development. As the technology advances, more children will be put at risk of serious harm. Governments and online platforms must take action to prevent the proliferation of these products.

Men's sexual entitlement to the bodies of women and children must be challenged, not accommodated. Rather than attempting to cater to men's sexual desires for children through the development of sexual outlets in their likeness, efforts need to be directed towards challenging the male sex right and the wider patriarchal culture that endorses and facilitates it. As I have outlined in this chapter, paedophilic men have historically employed a range of tactics to justify their sexual use of children, and now, child sex abuse dolls. Despite their best attempts at obfuscation, portraying child sex abuse dolls as the possible means of preventing child sexual abuse, their arguments remain unconvincing and transparent. Crimes of rape and child sexual abuse are not prevented with practice. Normalising men's sexual use and abuse of children might benefit some men, but it will never be in the best interests of children. Men's sexual preferences cannot come at the expense of children's human rights.

References

Angelides, S. 2004. Feminism, Child Sexual Abuse, and the Erasure of Child Sexuality. *GlQ: A Journal of lesbian and Gay Studies* 10(2), 141–177.

BBC Three. 2018. *The Future of Sex? Sex Robots and Us*. [online video] Available from https://www.youtube.com/watch?v=kGTI2_O9v_Y. Accessed 7 Sept 2021.

Behrendt, M. 2017. Reflections on Moral Challenges Posed by a Therapeutic Childlike Sexbot. International Conference on Love and Sex with Robots, 96–113. Springer, Cham.

Bender, L., and A. Blau. 1937. The Reaction of Children to Sexual Relations with Adults. *American Journal of Orthopsychiatry* 7 (1937): 514.

Berlatsky, N, 2021, Stigma Makes It More Difficult To fight CSA. *Prostasia*, 13 December. Available from https://prostasia.org/blog/stigma-makes-it-more-difficult-to-fight-csa/.

Bermingham, K. 2020. Shane Andrew Lunnay, Who Streamed Himself Having Sex with a Child-Like Sex Doll, Jailed for Importing Child Exploitation

Material. *news.com.au*, 23 September. Available from https://www.news.com.au/national/south-australia/shane-andrew-lunnay-who-streamed-himself-having-sex-with-a-childlike-sex-doll-jailed-for-importing-child-exploitation-material/news-story/77b2497ad47f68c7bf1ce342525c4ea6.

Brown, R., and J. Shelling 2019. Exploring the Implications of Child Sex Dolls. Trends and Issues in Crime and Criminal Justice. *Australian Institute of Criminology* 570, 1–13.

Califia, P. 1991. Feminism, Pedophilia, and Children's Rights. *Paidika* 2(4) reprinted in The Culture of Radical Sex (1994). Available from https://www.ipce.info/ipceweb/Library/califa_feminism.htm.

Califia. 1994. The Aftermath of the Great Kiddy-Porn Panic of '77'. *The Culture of Radical Sex*. Available from https://www.ipce.info/ipceweb/Library/califa_aoc_frame.htm.

Cantor, J. 2017. [Twitter] 5 February. Available from https://twitter.com/JamesCantorPhD/status/828009118736777216.

Cantor, J. 2018. [Twitter] 13 October. Available from https://twitter.com/JamesCantorPhD/status/1050816822189154304.

Cantor, J. 2019a. [Twitter] 6 December. Available from https://twitter.com/jamescantorphd/status/1202697473254400002?lang=en.

Cantor, J. 2019b. [Twitter], 22 September. Available from https://twitter.com/JamesCantorPhD/status/1175787768909377536.

Child Rescue Coalition. 2020. A Mother's Story: Daughter's Photo Stolen & Likeness Turned Into A Child Sex Doll. *Child Rescue Coalition*, 1 October. Available from https://childrescuecoalition.org/educations/child-sex-doll.

Chung, L. 2021. Man Charged Over Alleged Importation of Childlike Sex Dolls. *The Age*, 10 September. Available from https://www.theage.com.au/national/nsw/man-charged-over-alleged-importation-of-childlike-sex-dolls-20210907-p58pjv.html.

Clásicas y Modernas. 2021, Pensar el Feminismo 15 mañana. [online video] Available from https://www.youtube.com/watch?v=5Ejz59OKQ04. Accessed 7 Sept 2021.

Collective Shout. 2013. Win! Bookworld Forced to Withdraw Incest Titles. *Collective Shout*, 6 February. Available from https://www.collectiveshout.org/win_bookworld_forced_to_withdraw_incest_titles.

Collective Shout. 2020. Collective Shout Welcomes Alibaba's Removal of Child Sex abuse Doll Listings. *Collective Shout*, 22 July. Available from https://www.collectiveshout.org/alibaba_removes_dolls.

Collective Shout. 2021. Media Release: Strong Action Against Sex Dolls by e-commerce Giant Alibaba. *Collective Shout*, 17 November. Available from https://www.collectiveshout.org/mediarelease_alibaba_geoblock.

Dines, G. 2011. The New Lolita: Pornography and the Sexualization of Childhood. In *Big Porn Inc: Exposing the Harms of the Global Pornography Industry*, eds. A. Bray and M. Tankard. Reist Spinifex Press.

Earp, B. 2017. Pedophilia and Child Sexual Abuse Are Two Different Things—Confusing Them is Harmful to Children. Blog. *Journal of Medical Ethics*, 11 November. Available from https://blogs.bmj.com/medical-ethics/2017/11/11/pedophilia-and-child-sexual-abuse-are-two-different-things-confusing-them-is-harmful-to-children/.

Elchuk, D.L., I.V. McPhail, and M.E. Olver 2021. Stigma-Related Stress, Complex Correlates of Disclosure, Mental Health, and Loneliness in Minor-Attracted People. *Stigma and Health* 2(1), 1–48.

Férdeline, A. 2020. Facts About Prostasia. Available from https://www.prostasia.info/. Accessed 7 Sept 2021.

Gilmore, J. 2017. Domestic Violence Half-Yearly Update: The Numbers We Need to Know. *Sydney Morning Herald*, 5 July. Available from https://www.smh.com.au/lifestyle/domestic-violence-halfyearly-update-the-numbers-we-need-to-know-20170704-gx4dsv.html.

Hall, J. 2019. 'Against her will': Amazon Sells Exploitative Books Glorifying Rape and Sexualising Children. *news.com.au*, 5 April. Available from https://www.news.com.au/finance/business/retail/against-her-will-amazon-sells-exploitative-books-glorifying-rape-and-sexualising-children/news-story/6e56c774e3504dcc1c6bafcf3835e572.

Hall, J. 2020. Etsy Seller Providing Childlike Sex Dolls Modelled Off 14-Year-Old Instagram Star. *news.com.au*, 9 December. Available from https://www.news.com.au/technology/online/etsy-seller-providing-childlike-sex-dolls-modelled-off-14yearold-instagram-star/news-story/5baac0a4b8c02dd3dd711d57e1bd9531.

Harper, C.A., and R. Lievesley 2022. Exploring the Ownership of Child-Like Sex Dolls. Arch Sex Behav. https://doi.org/10.1007/s10508-022-02422-4.

Harper, C.A., R. Lievesley, N.J. Blagden, and K. Hocken 2021. Humanizing Pedophilia as Stigma Reduction: A Large-Scale Intervention Study. *Archives of Sexual Behavior*, *51*(2), 1–16.

Henderson, D.J. 1975. Incest. In *Comprehensive Textbook of Psychiatry*, 2nd ed., eds. A.M. Freedman, H.I. Kaplan, and B.J SadockThe Williams & Wilkins Company.

Herman, J. 1981. *Father Daughter Incest*. Cambridge MA: Harvard University Press.

Hetherington, G. 2019. Former Soldier Found with Child-Like Sex Doll Jailed for Sex Crimes. *The Northern Echo*, 13 August. Available from https://www.thenorthernecho.co.uk/news/17832017.former-soldier-found-child-like-sex-doll-jailed-sex-crimes/.

Hobday, L. 2021. Paedophile Primary Teacher Uses School Photos of His Students to Make Sick Videos of Himself with a Childlike Sex Doll Dressed in Uniform, Daily Mail, 8 November. Available from https://www.dailymail.co.uk/news/article-10176409/James-William-Treasure-Paedophile-primary-teacher-dresses-sex-doll-school-uniform.html.

James, J. 1970. *The Facts of Sex*. Princeton, NJ: Vertex.

Kennedy, L. 2019. 'Sexy girl': How Instagram Allows the Offering of Young Girls as Fetishised Flesh. *Collective Shout*, 13 September. Available from https://www.collectiveshout.org/sexy_girl_instagram_allows_young_girls_fetishised.

Levy, D. 2007. *Love and Sex with Robots: The Evolution of Human–Robot Relationships*. New York: Harper.

Liszewski, M. 2018. Child Sex Dolls Removed from Online Store Wish. *Collective Shout*, 30 October. Available from https://www.collectiveshout.org/child_sex_dolls_removed_from_online_store_wish.

Lockley, K., and E. Barr 2021. Steven Capsis Charged with Fresh Possession of Child Porn Offences. *Daily Telegraph*, 5 May. Available from https://www.dailytelegraph.com.au/newslocal/stgeorge-shire-standard/steven-capsis-charged-with-fresh-possession-of-child-porn-offences/news-story/122eb092b908ecde1f21bc68ed21fb87?btr=4732472f5181977811ea4c54_2fe0d74.

Malcolm, J. 2021a. [Twitter] 18 June. Available from https://twitter.com/qirtaiba/status/1405715180277813248.

Malcolm, J. 2021b. [Twitter] 24 June. Available from https://twitter.com/qirtaiba/status/1407865086664794114.

Malcolm, J. 2021c. [Twitter] 24 June. Available from https://twitter.com/qirtaiba/status/1407754968782688256.

Malcolm, J. 2021d. [Twitter] 24 June. Available from https://twitter.com/qirtaiba/status/1407851020919603204.

Malcolm, J. 2021e. [Twitter] 5 November. Available from https://twitter.com/qirtaiba/status/1456649544700088320.

Moen, O.M., and A.B. Sterri 2018. Pedophilia and Computer-Generated Child Pornography. In *The Palgrave Handbook of Philosophy and Public Policy*, ed. D. Boonin. Palgrave Macmillan.

Morin, R. 2016. Can Child Dolls Keep Pedophiles from Offending? *The Atlantic*, 11 January. Available from https://www.theatlantic.com/health/archive/2016/01/can-child-dolls-keep-pedophiles-from-offending/423324/.

Mott, M. 2021. Adam Neil Wonnocott Jailed for Three Years and Six Months for Possessing Thousands of Child Exploitation Images. *The Advertiser*, 28 May. Available from https://www.adelaidenow.com.au/truecrimeaustralia/police-courts-sa/adam-neil-wonnocott-jailed-for-three-years-and-six-months-for-possessing-thousands-of-child-exploitation-images/news-story/ff4c0edf00d713b82e27900979243700?btr=462db2dc176853e063a4cb352c0805a6.

Nicol, R. 2019. Governor Signs Bill Outlawing Child-Like Sex Dolls. *Florida Politics*, 23 May. Available from https://floridapolitics.com/archives/297216-governor-signs-bill-sex-dolls/.

Pegg, S. 2016. Could Your Taste for 'teen' Porn Land You in Legal Trouble? *The Conversation*, 15 January. Available from https://theconversation.com/could-your-taste-for-teen-porn-land-you-in-legal-trouble-53179.

Prostasia Foundation. 2019a. Our Campaigns Against Doll Bans. *Prostasia Foundation*. Available from https://prostasia.org/sex-doll-laws/.

Prostasia Foundation. 2019b. [Twitter] 3 January. Available from https://twitter.com/ProstasiaInc/status/1080518088687267841.

Prostasia Foundation. 2021. [Twitter] 16 February. Available from https://twitter.com/ProstasiaInc/status/1361514207242383360.

Prostasia Foundation. n.d. Exploring the Use of Fictional and Fantasy Sexual Outlets. *Prostasia Foundation*. Available from https://prostasia.org/project/research-fund/.

RAINN. 2022. Child Sexual Abuse. *Rape, Abuse & Incest National Network*, n.d. Available from https://www.rainn.org/articles/child-sexual-abuse.

Ramey, J. 1979. Dealing with the Last Taboo. *SIECUS Report 7*, May, VII(5), 1–16.

Rawlinson, K. 2017. Child Sex Doll Imports Expose Previously Unknown Offenders. *The Guardian*, 31 July. Available from https://amp.theguardian.com/society/2017/jul/31/child-sex-doll-imports-expose-previously-unknown-offenders?CMP=share_btn_tw&__twitter_impression=true.

Reavy, P. 2021a. Disturbing Find Made During Child Porn Investigation. *KSL.com*, 8 September. Available from https://www.ksl.com/article/50237290/disturbing-find-made-during-child-porn-investigation.

Reavy, P. 2021b. Orem Man Charged in Child Porn Case After Police Report a Disturbing Find. *KSL.com*, 17 May. Available from https://www.ksl.com/article/50108546/orem-man-charged-in-child-porn-case-after-police-report-a-disturbing-find.

Reisman, J., and M. McAlister 2018. Deconstructing Dignity by Eradicating Shame: The Pernicious Heritage of Alfred Kinsey. *Faculty Publications and Presentations*. 4. https://digitalcommons.liberty.edu/psych_fac_pubs/4.

Roper, C. 2015. Buyinvite Exposed Selling Child Sexual Abuse Themed Sex Toys for Men. *Collective Shout*, 23 July. Available from https://www.collectiveshout.org/buyinvite_exposed_selling_child_sexual_abuse_themed_sex_toys_for_men.

Roper, C. 2020a. "Better a robot than a real child": The Spurious Logic Used to Justify Child Sex Dolls. ABC Religion and Ethics, 9 January. Available from https://www.abc.net.au/religion/spurious-logic-used-to-justify-child-sex-dolls/11856284.

Roper, C. 2020b. Not a 'Victimless' Crime: How Child Sex Abuse Dolls Facilitate Crimes Against Children. *FILIA*, 10 September. Available from https://www.filia.org.uk/latest-news/2020b/9/10/not-a-victimless-crime-how-child-sex-abuse-dolls-facilitate-crimes-against-children.

Roper, C. 2022. Etsy Ignores Calls to Remove Child Abuse Merch: It's been a Year But We're Not Stopping! *Collective Shout* 2 February. Available from https://www.collectiveshout.org/etsy_ignores_calls_to_remove_child_abuse_merch.

Rubin, G. 1984. Thinking Sex: Notes for a Radical Theory of the Politics of Sexuality. In *Culture, Society & Sexuality*, ed. R. Parker, and P. Aggleton, 1998, Taylor & Francis Limited.

Rush, F. 1980. *The Best Kept Secret: Sexual Abuse of Children*. New Jersey: Prentice-Hall.

Seto, M.C. 2012. Is Pedophilia a Sexual Orientation? *Archives of Sexual Behavior* 41 (1): 231–236.

Slatz, A. 2021. Prostasia's Goal is to Normalize Pedophilia. *4W*, 29 August. Available from https://4w.pub/prostasia-normalize-pedophilia/.

Tankard Reist, M. 2008. Incensed About Censorship. *ABC*, 27 November. Available from https://www.abc.net.au/news/2008-11-27/28374.

UN Special Rapporteur. 2020. Report of the Special Rapporteur on the Sale and Sexual Exploitation of Children, Including Child Prostitution, Child Pornography and Other Child Sexual Abuse Material. *Human Rights Council Forty Third session*, no. 40, 24 February–20 March. Available from https://undocs.org/en/A/HRC/43/40.

Virped. 2012. Welcome to Virtuous Pedophiles. Virped. Available from www.virped.org. Accessed 10 Oct 2021.

Walker, A. 2021. *A Long*. Dark Shadow: Minor-Attracted People and Their Pursuit of Dignity, University of California Press.

Wiseman, E. 2015. Sex, Love and Robots: Is This the End of Intimacy? *The Guardian*, 13 December. Available from https://www.theguardian.com/technology/2015/dec/13/sex-love-and-robots-the-end-of-intimacy.

World Health Organisation. 2003. Child Sexual Abuse. Guidelines for Medicolegal Care for Victims of Sexual Violence. Available from https://apps.who.int/iris/bitstream/handle/10665/42788/924154628X.pdf.

World Population Review. 2022. Age of Consent by Country 2022. *World Population Review*. Available from https://worldpopulationreview.com/country-rankings/age-of-consent-by-country.

CHAPTER 8

The Voice of the 'Sex Robot': From Peep-Show Bucket to Willing Victim—The Terrorism of Women's Speech

Shirley MacWilliam

But the sex doll is meant to be as selfless as it is willing, always and fully open to the boldest advance. If it speaks, it must vent on command, crying out in phonographic orgasm when a man presses her lips with kisses.

Hillel Schwartz *The Culture of the Copy* (127).

In the history of culture, woman has been denied the function of speech.

Susanne Kappeler *The Pornography of Representation* (65).

S. MacWilliam (✉)
Belfast School of Art, Ulster University, Belfast, UK

© The Author(s), under exclusive license to Springer Nature Switzerland AG 2022

K. Richardson and C. Odlind (eds.), *Man-Made Women*, Social and Cultural Studies of Robots and AI,
https://doi.org/10.1007/978-3-031-19381-1_8

This chapter will explore the voice of the 'sex robot' and argue that, while presented under the guise of narratives of love, technological advancement and art, the voice of the sex robot encapsulates a pornographic illustration of woman as willing sexual victim. I propose that the voice of the sex robot is a physical manifestation of a culture of onslaught upon women's voices, and a pattern, during the past century, of appropriation and reframing of women's speech by art, literature, technology, sex industries and psychoanalysis. Women's voices are sexually objectified to sexually harass women, to enforce audible expressions of sexual pleasure and to interfere with the public presence of women's political speech.

I will introduce the idea of the sex robot, its 'voice' and capacity for 'love.' Then I will explore examples of women's voices in decadent art, the suffragette movement and psychoanalysis in the late nineteenth century and early twentieth century. Using Susanne Kappeler's radical feminist analysis of the 'peep-show bucket' and the scenario of the willing victim, and with reference to the writing of feminists Sheila Jeffreys and Kajsa Ekis Ekman, I will trace a consistent misogynist trajectory from early voice recordings through to the present day. I will show how the 'voiced' sex doll is one in a long line of technologies bringing 'voice' pornography and 'porn culture' into public space.

Love and the Sex Robot

The 'sex robot' is a fiction.[1] It is not a technical reality, it is a cultural idea which sells news, films, advertising, sex dolls and pornography. The reality is that many companies internationally make and sell dolls with a variety of 'animating' features (such as temperature controls and moaning sounds) and a few individuals, teams or laboratories claim to be developing more elaborate integrations of artificial intelligence (AI).

The primary AI feature which is purported to 'transform' the 'sex doll' into a 'sex robot' is a 'voice.' Accompanied by head and facial speech-related movement, the voice of a Realbotix robot, for example, is controlled by a mobile phone app which represents the doll as an avatar 'girlfriend.' Selling points are that the user can "Choose a voice, shape a personality," and that the doll will speak when spoken to and can be turned off. Additionally, a tactile sensor device located in the synthetic

[1] Kathleen Richardson's clarity about the fictional status of the robot is instructive.

'vaginal' orifice of the doll connects to the synthetic voice.[2] The voice will respond to stimulus by expressing encouragement and producing sounds representing sexual arousal.

Realdolls already market 'love' dolls rather than 'sex' dolls, however, the development of the AI voice, and thereby 'robot' status, introduces a new dimension of 'love story' to the product. Companionship, personality, love and intimacy are promised with and by the voice.

Several contemporary philosophers have turned their attention to the promise of robot love. Sven Nyholm and Lily Eva Frank, for example, propose that we can consider 'mutual love' with a sex robot possible on the basis that 'in principle' a manufactured object with sophisticated AI could fulfil several criteria, or as they put it 'a job description.' for 'mutual love.' Their approach focuses on "clusters of ideas about what people typically seek and value in romantic love that can be found, not only in the philosophy of love, but also in art and literature, pop culture, and in everyday thinking about love" (Nyholm and Frank 2017: 220). Nyholm and Frank, however, take these clusters of ideas as a barometer of love without considering the practices, values and sexual politics within which they are shaped.

Thus, if love is considered in relation to "being a good match," "being made for each other," "arranged marriages," "Eve… created from Adam's rib," "divine plan" or in relation to "Aristotle's classic phrase 'one soul in two bodies,'" they argue that love with the robot is credible. They suggest that the possibility of 'love' might be advanced by having to "woo the robot," "to win its love," presumably as one plays a computer game (227–229).

The essay includes wishful, tautological formulations such as "If a robot were equipped with the capacity to come to love humans romantically…" and "It would need to be a robot to which we can sensibly ascribe the ability to fall in love" (229). They claim that the costly pursuit of these fictions is worthwhile, in the name of love.

Most of their references suggest patriarchal conceptions of love and some of their 'criteria' are techniques evident in abusive relationships, such as the capacity to mimic empathy and love behaviours in the absence of any feeling, and the emotionally manipulative insistence: "It had to be

[2] In an odd sex-tech vagina dentata conflation, the device is called: SenseX Bluetooth.

you."[3] They gesture to dominant sexist and market driven clichés about love and extoll love robots on the logic that: love is good, a love robot produces love, therefore love robots are good. Tellingly, their statement "First you might buy the robot…," and the centrality to their analysis of a "job description," remain unexamined. To buy or employ one's 'love object' and therefore to forge a 'relationship' based on purchase or employment is far from mutuality and love and clearly more like slavery and prostitution. The proofs of love cited are the very conventions (or clusters of ideas) which feminists have shown to be designed to enshrine the subordination of women (Beauvoir [1949] 1953, Millet [1969] 1977).

No mention is made that sex doll owners are almost universally male[4] and sex dolls/robots almost universally represent women and girls. Ultimately, this reflection on love (which might be summarised as *love makes sex better, therefore, love robots are sexier than sex robots*) is philosophical window dressing for the sex-tech industry.

A Decadent Love Story: He Took Her Breath Away

In 1891, Marcel Schwob, a twenty-four-year-old man living in Paris, published a 'love story' called *Beatrice*. The name Beatrice refers to a popular nineteenth-century literary archetype of ideal love which harks back to Dante. Dante's literary Beatrice was based on a real thirteenth-century woman, whom he met as a child and who died young. Dante's writing created the archetypal Beatrice who is exquisite, blissful, a channel for transcendent truth, a spirit guide and a dead woman (Hall [1974] 1983: 42–43). Beatrice died in the thirteenth century and dies over and over again in art. Schwob was a writer and translator and part of the fin-de-siècle French and English decadent movement in literature and art. In his version of *Beatrice*, he describes an intense antipathy towards a woman's voice and invokes love values like those found in discussions about automated machines shaped like women.

[3] *It had to be you* is also the opening title music of *When Harry Met Sally* which I discuss later.

[4] It is a sex industry marketing success that several female artists have bought sex dolls (or been sponsored) and made artworks and photography featuring them which are then exhibited, published and discussed across a wide range of public platforms including the social media of the manufacturers.

Sheila Jeffreys describes the European fin-de-siècle period as a "watershed in the history of human sexuality" involving sexual revolution, social purity movements and "a massive campaign by women to transform male sexual behaviour and protect women" (Jeffreys 1985: 1). It is also the era of the rise of industrial pornography.

In the 1880s, "a 'pornographic crisis' erupted in France as a surge in cheap lubricious periodicals inundated the streets of Paris" (Rexer 2021: xi). In previous decades, printing methods of the photograph, the postcard and the stereoscope had enabled the first industrial production and mass distribution of visual pornography. Art historian Gerald Needham notes the "immense production" of "pornographic photographs ... widely distributed in the major cities of Europe." For example, "The [Saturday] *Review* [in 1858] complained that 'there is hardly a street in London which does not contain shops in which pornographic photographs and especially stereoscopic photographs are exposed for sale'" (Needham 1972: 81).

Marcel Schwob moved to Paris to attend school when he was fourteen in 1881. Thus, he is among the first generations of boys to have grown up exposed in the public realm to easily accessible mass-produced pornographic images of women.

Schwob was also influenced by other technological developments. In 1889 his father, who owned a newspaper, published an interview with Thomas Edison who discussed his desire to develop the commercial application of the phonograph. The following year on the anniversary of poet Robert Browning's death a group of Browning's admirers gathered to listen to an Edison recording of Browning's voice. The event was widely reported in literary circles and described as the first voice heard from beyond the grave (Green 1965).

The decadent movement of which Schwob was a part was sexually libertarian. The artists and writers cast themselves as refined, sophisticated and melancholic and they celebrated ideas about art as transgressive, sexually experimental and opposed to social conservatism. The same type of ideas flourish today in the culture, funding and promotion of contemporary art.

The movement was fascinated by artifice, corruption and pleasure-seeking. Decadent artists and writers eroticised sickness, debasement and aligned love and sex with death, decay and venereal disease. Schwob and other artists and writers were enthralled, in their lives and art, by prostitution and particularly the 'child-prostitute.' Women were represented as either pure, nervous, anaemic, wasting away or pox-ridden, ghoulish and sexually compelling. In the decadent world virgins, whores and artists die. Syphilis was endemic across Europe and much discussed, moralised and feared. The decadent "literary mythology" and "fantasies" of venereal disease (Showalter 1990) amounted to a cult of syphilis: a love-sex-death cult.

In the story *Beatrice* Schwob combines ancient and contemporary myths of love with the idea, made real by the phonographic technology of his day, that a voice can be heard after death.

The character Beatrice was known to the narrator-protagonist since she was a child (and younger than him) and is idealised by him as a work of art. Her appearance is described in the language of sculpture: "her skin as white as marble; she was a piece of statuary"; "cleanly and sharply cut"; "a figure of alabaster… illuminated from within by a rosy lamp"; "Beatrice was a marble statue" (Schwob [1891] 1992: 282–286). As his love 'object,' she is never fully human, she is the man-made creation of Schwob, of the narrator, of "the master-sculptor himself" (283).

The narrator describes his fascination with the Platonic conception of love as absolute union in the exchange of souls and he says that he and Beatrice have an implicit pact to achieve it. This entails, however, not a mutual bonding but Beatrice's unspoken acquiescence to make a "sacrificial offering" by letting her life drain away. He watches Beatrice waste and wane, increasingly pale, anaemic and emaciated: "I knew in the end, that Beatrice must belong to me entirely within a few days – and in spite of my infinite sadness, a mysterious joy extended itself within me." Beatrice has one line of speech, and it is her last: "At the moment when I die, I wish that you should kiss me on the mouth and that my last breath should pass into you" (286).

The idea of love as the meeting of soul mates (the mutual intermingling of souls or of two half souls reunited) is replaced by the practice of love which Simone de Beauvoir describes in terms of man's sovereignty and woman's abandonment: "she lets her own world collapse in contingence, for she really lives in his"; "she is another incarnation of her loved one, his reflection, his double: she is *he*" (Beauvoir [1949] 1953: 618).

At her death—the moment of love union—Beatrice has "the appearance of a statue of virginal wax." As the narrator inhales her last breath from her body he intends to gain her spirit, her being, "to drink her soul." He is appalled to discover that he has no feeling of her soul within him, but that he now speaks in her voice. "I had killed Beatrice and I had killed my own voice; the voice of Beatrice dwelt in me now" (Schwob [1891] 1992: 286).

The narrator remains able to narrate his story even though he has Beatrice's voice. However, losing his own voice and having to listen to the sound of hers is intolerable to him. His story is primarily an explanation of why he is killing himself. It is framed by him lying in perfumed water having cut the wrist of one hand while he uses the other to write his tale. His grim egotism is demonstrated by the fact that the servants, who hurry to attend to the dead woman, do not perceive that he has her voice. He possesses it via his hallucination—his relentless projection of his version of reality onto the world.

Beatrice has been turned into a dead body object, like the artwork she always was, and her voice taken from her. Human voice is formed at the meeting points of body and language, inside and outside personal and social. Voice is composed of the same air we need to live and breathe. To have a voice is to have a say, to have social status. To realise you are voiceless is terrifying.

The Voice, Vote and Sexual Politics of the Suffragettes

> This cycle of women's voices bursting forth only to be interrupted and silenced [...] *is* women's tradition. This censoring and suppression of women's words of protest, time after time, is the common heritage of women. (Spender 1983: 3)

Although sometimes described as first wave feminists, the turn of the century suffragettes were not the first to give voice to the social and sexual disadvantages accorded to women, as Dale Spender attests. However, while the decadent art movement was ventriloquising dying women and revelling in its cult of syphilis, these women were not only arguing for the vote but also publicly articulating how male sexual behaviour throughout society was damaging women and girls, inside and outside the home. Spender (1983) and Jeffreys (1985, 1987) show how, alongside the fight for political enfranchisement and improvements in educational and economic status, the first wave women's movement developed a radical feminist analysis of male sexuality as a structure of social control of women. Women found the language and the public voice to describe and challenge male sexual behaviours: prostitution, the sexual abuse of children, the Contagious Diseases Act, the marriage contract, domestic violence, rape and public sexual harassment.

Suffragette sexual politics directly countered dominant patriarchal theories such as those of London doctor William Acton, who held prostitutes responsible for the moral and venereal corruption of men and blamed women for men's sexual objectification of women. He proposed that using a prostituted woman would lead a man "to make, if possible, of every woman the thing that he desires – a toy, a plaything, an animated doll, a thing to wear like a glove and fling away; to use like a horse and send to the knackers when worn out" (in Jeffreys 1987: 43). According to Acton, it was women's fault that men wanted to turn them all into 'animated dolls.'

The feminist analysis showed how dividing women into the pure and polluted was part of the method of the sexual subordination of all women. Christabel Pankhurst, for example, connects the permanent sex bargain of the wife with the temporary one of the prostitute. The feminists explained how women were blamed and punished for the spread of sexual diseases by men. Under the Contagious Diseases Act, women, suspected of prostitution were imprisoned in 'lock hospitals' where they were subjected to sexual inspections and treatments until they were considered 'clean' of venereal disease, and therefore suitable again for sexual use and reinfection by men. Women were treated, by the state, as a facility to be regularly flushed out for men's benefit. The feminists also revealed how married women were frequently infected with venereal diseases by their husbands but rarely had any knowledge that this was the case. It is deeply telling and shocking that so many syphilitic ailments and symptoms experienced by women are the same as the characteristics ascribed to Victorian femininity: anaemia, consumption, epilepsy, insanity, miscarriage, nervousness, neuralgia, blindness, paralysis, hysteria. The consequences of male promiscuity, suffered by women, were disguised as the 'nature' of the 'weaker sex.'

As well as courageous truth-telling and campaign for legal reform, the suffragettes advocated radical feminist responses: women might say 'no'; they might choose not to marry; they might choose chastity (and make a political decision not to have sexual relationships with men); they might prioritise their love of women above a duty to men. Equally, they argued, men were responsible for their sexual behaviours, and they should change them.

Christabel Pankhurst asked directly that men keep their "desires under control" but this interference with male sovereignty threatened that special domain of its expression, art. In response, the artist-writer makes: "his protest in the name of art and asks indignantly 'Do you think that any artistic manifestation could come out of chastity and normality?'" (Pankhurst [1913] 1987: 335). In *The Picture of Dorian Gray*, Oscar Wilde's character Henry Wotton voices the same sentiment—that women with opinions and lived lives impede the artist: "Women inspire us with the desire to do masterpieces and always prevent us from carrying them out" (Showalter 1990: 176). Women's attempts to exert boundaries and prevent sexual exploitation are cast as the enemy of art and artists.

The tactics for silencing and discrediting the women's movement were varied and sustained. Suffragettes were ridiculed and reviled in popular culture. In person, when they spoke or gathered to protest, they were shouted down, physically stoned, charged by mounted police, violently attacked by men in gangs and subjected to orchestrated sexual assault by policemen. In jail, they were choked and brutalised by force-feeding. In literature, feminist speakers were described as mannish, prudes, repugnant, out of place, immodest, odd, vulgar, a "shrieking sisterhood." Novelist Eliza Lynn Linton describes a female orator "whose case-hardened self-sufficiency was as ugly as a physical deformity"; Henry James presents "'a confused, entangled, inconsequent, discursive old woman' a monitory image of what befalls women who take up public speech" (Showalter 1990: 23–24). It was a concerted backlash on many fronts. The vote was achieved but the feminist account of sexual politics was suppressed.

Jeffreys records how the rise of sexology (the 'science' of sex) and the early twentieth-century sexual revolution silenced women's voices and reinforced the sexual objectification of their bodies: "The struggle of women to assert their right to say no gradually faded into insignificance whilst male sex theorists debated astride the conquered territories of women's bodies" (Jeffreys 1985: 5).

Hysterical Mendicants, Hysterical Bleeders and Blueprints for Pornography

Psychoanalysis, born in the era of the suffragettes and decadents, forms another vanguard which purposefully silenced women's voices that spoke about male sexual violence.

Like the women's movement, the earliest foundations of psychoanalysis were an exceptional opportunity for the testimony of women's experience. In the context of the 'talking-cure,' women spoke about childhood sexual abuse, usually by their fathers, and they were heard. Initially, Sigmund Freud listened and in *The Aetiology of Hysteria* (1896) he formulated his radical theory of seduction: that many neuroses experienced by adult women arose from their experiences of rape and sexual abuse as infants and young girls. Over several years, however, Freud reframed what they told him as psychical fantasy, and he based his theory of sexual development on his disavowal of the women's speech. Psychologist Judith Herman places this in its political context:

> To hold fast to his [first] theory would have been to recognise the depth of the sexual oppression of women and children. The only potential source of intellectual validation and support for this position was the nascent feminist movement, which threatened Freud's cherished patriarchal values. To ally himself with such a movement was unthinkable for a man of Freud's political beliefs and professional ambitions. Protesting too much, he dissociated himself at once from the study of psychological trauma and from women. He went on to develop a theory of human development in which the inferiority and mendacity of women are fundamental points of doctrine. In an antifeminist political climate, this theory prospered and thrived. ([1992] 1994: 19)

In *The Assault on Truth* Jeffrey Masson explores the papers and correspondence which document this process. He confirms Herman's view that, were Freud to have endorsed the women, it would have been socially and professionally catastrophic. Masson shows, however, that Freud was for some time troubled by the "clinical evidence" of "frightening explosive scenes which can affect the victim's entire life" (Masson 1992: 117) and by his conviction that fathers often had sexually abused their daughters (92). Most startling though, is the role of Freud's infatuation with fellow doctor Wilhelm Fliess and his consequent incentive to prioritise an invented 'psychical reality' above material reality. For over a decade, Freud and Fliess corresponded intensely, sharing discoveries, theories and information about patients, including the case and treatment of Emma Eckstein. Masson explains that Eckstein was active in the women's movement in Vienna and entered analysis with Freud due to physical lassitude, stomach and menstrual pain.

Fliess cultivated medically tenuous theories about numerology, biorhythms, relationships between sex organs and the nose, and cocaine centred remedies. He suggested to Freud that he operate on Eckstein's nose. Fliess proposed to cut out a piece of bone to cure a presumed masturbation problem. Despite his extreme scepticism about Fliess' naso-genital theories, Freud agreed. Given the experimental nature of both their practices, it is hard not to see Emma Eckstein as anything other than a convenient woman exchanged between them, like a laboratory specimen.

Back in Freud's care after surgery, Eckstein suffered two major life-threatening haemorrhages, which occurred because Fliess incompetently left a long piece of gauze in her nose. Extraordinarily, even though the gauze was acknowledged between them, Freud, in his infatuation, was incapable of holding Fliess responsible. Rather than confront Fliess, Freud "sought the cause of the bleeding in Emma Eckstein herself" (99). He constructed "a diagnosis of hysterical bleeding" (72) and he wrote that: "she bled out of *longing*. She's always been a bleeder [...] she renewed the bleedings, as an unfailing means of rearousing my affection" (101), "her haemorrhages were due to wishes" (102).

Masson imagines Freud's thought process:

> To efface the external trauma of the operation, it would prove necessary to construct a theory of hysterical lying, a theory whereby the external traumas suffered by the patient never happened but are fantasies. If Emma Eckstein's problems (her bleeding) had nothing to do with the real world (Fliess' operation) then her earlier accounts of seduction could well be fantasy too. (99)

Freud conceals the crimes, vanity and incompetence of men with his newfound axiom: "the mendacity of hysterics" (186).

Sexual exploitation has a historical evolution. In its first phase, a patriarch can do whatever he wants to those below him, he is officially sanctioned, and exploitation is known and accepted. In the second phase, there is official denial that these things happen: exploitation is hidden. In the third phase, the facts of exploitation emerge and psychoanalytic revisionism (or some other ideology) produces an ideological alibi which construes facts as fantasy and injury as invited.

Using women's voices, Freud took the reality of child rape and sexual abuse and represented it to the world as fantasy within a carefully formed 'scientific' theory. The thoughtful excavation of memories which once he believed uncovered the hidden truth of real experiences was replaced by a mystical, mysterious and wonderful structure: the unconscious. The story he told in the theory of the unconscious, which is primarily a theory designed to accommodate male sexuality, was that one cannot know one's own experiences or feelings, but that the unconscious is all knowing and only indirectly accessible. The mythic unconscious resonates with superstition and intellectual grandiosity and self-certifies that 'truth' is too complex and astonishing to be available to the prosaic conscious mind.

Like the lies told to women about themselves to explain the impact of syphilis, Freud tells women that their memories, reactions to and symptoms of sexual violence are really their innate fantasised desires; they are their femininity. He produces the concept of the "sexual constitution" of the girl: "The 'neurotic' adolescent does not want to acknowledge her own sexual desires and in order to cover them up she 'invents' sexual tales of seduction from her early childhood" (122).

By suggesting women cannot trust themselves (and cannot be trusted), Freud casts doubt on the protective touchstones, 'say what you see' and 'listen to your gut,' the basic human safety mechanisms of direct sense perception, and subconscious instinct.

The dexterity of Freud's theory is that it denies the reality of sexual abuse and simultaneously plants the idea of the sexually motivated child. If it were discovered that an adult had done something sexual to a child it became plausible to interpret it as something the child wanted, desired or initiated. Freud theoretically sexualised the infant because the infant had been physically sexualised by the abuser. To cover and repudiate the abuse, he produced the idea of infantile sexuality, in the form of the sexual girl child, who wants to seduce an adult man. The silence demanded of a child by an abuser was perpetuated by the analyst.

To silence women, he produced the idea of woman as 'hysterical mendicant' and 'hysterical bleeder' (neither the evidence of her voice nor of her body can be believed); the woman whose desires are unknown to her because they are repressed; the woman whose revulsion is evidence of her repressed desire.

The intricate infrastructure of Freud's story of sexual development (the foundation story of psychoanalysis) reproduces the misogynist principles upon which sexology, the sexual revolution and everyday relationship psychology are founded. Furthermore, Freud validates the blueprints of contemporary pornography: the woman who wants everything to which she says 'no'; the sexually available seductive child; and even the casual designation of the 'daddy.'

The Peep-Show and the Peep-Show Bucket

In the year *Beatrice* was published, Edison's Kinetoscope was invented. The Kinetoscope was a proto-cinematic technology which showed a short length of film to an individual viewer. It was succeeded by the more commercially practical Mutoscope, a coin and crank operated free standing peep-show box with a flick-card reel, "above all famous for the 'Girlies' dance and strip-tease subjects" (MOMA 1967: 2). These mechanical peep-shows brought photographic, moving-image pornography, among other subject matters, into public spaces: men's public toilets, waiting rooms, piers and penny amusement arcades. Many of these locations were frequented by women, who thus walked past and stood beside men, knowing that they were watching sexually titillating, 'what-the-butler-saw' scenarios. The peep-shows invited men to exert a public sex right by choosing to voyeuristically objectify women in full sight of others.

To illustrate her proposition that, in the realm of pornography, *viewing* is *doing*, Susanne Kappeler draws attention to the peep-show bucket. She writes: "The ultimate symptom of this are the hundreds of thousands of men ejaculating into a bucket in their booths at the peep-shows" (1986: 58).

> It is not difficult to see why women fear that the doing might extend even further. It is not difficult to envisage a continuing process in the 'adult business' of live peep-shows to a point where instead of a bucket some other 'object' might be supplied. (59)

It does not escape women's notice that almost every new technological development becomes exploited in pornography and the wider sex industry. The live peep-shows of the 1970s and 1980s have 'evolved' in an explosion of technological 'sex show' options: cam girls and OnlyFans, hidden surveillance cameras, up-skirting, deepfakes, online 'menus' of prostituted women, hook-up culture and soft porn 'norms' throughout popular mainstream media. The sex doll/robot is a high-spec peep-show bucket: an elaborate masturbation receptacle incorporating visual cues from the 'live sex show.'

The contemporary live sex show is everywhere: in the acceptance of prostitution practices as normal everyday sexuality (Barry 1995); in the culture of 'sex work' as work; in franchised lap-dancing businesses and night clubs encouraging female customers to compete as the sexiest stage performer; and in "the cult of the whore" (Ekman 2013: 34)—the casual widespread adoption of the word 'whore' by women as a fashionable supposedly stigma-challenging epithet applicable in any circumstance. Common usage includes book whore, fashion whore, music whore, dance whore. "The whore is quintessentially hip" (34) and visible everywhere

Maintenancer (2018), a documentary film made by artists Sidsel Meineche Hansen and Therese Henningsen and exhibited in the Venice Biennale 2022, encapsulates the sex doll as bucket. It displays how dolls in a brothel are cleaned, between use by different men: "... she thoroughly cleans each of its orifices using wet wipes, inserting a black gloved hand in order to completely sterilise the doll" (Kilby 2020). The brothel owner, a 'dominatrix,' insists that she is running a regular brothel not a doll brothel. She refers jokingly to the dolls as ideal workers and employs another woman as 'the maintenancer.' As well as cleaning and preparing the dolls, this woman deals with the buyers. She 'cares' for the dolls and 'cares' for the men, proposes writer Benison Kilby. The narrative of the film and of Kilby's commentary is that 'sex work' is 'care work' and undervalued because it is done by women.

Of the sex show viewer Kappeler says: "what he is doing is watching, imagining, fantasising, producing the feeling of life in himself" (1986: 59). The idea of 'the feeling of life' comes from Kant's notion of the aesthetic experience of the art spectator who is untroubled by responsibility to the realities of the subject matter beyond the artwork. Thus in "the pursuit of pure self-interest" (56) the spectator experiences a feeling of animation and life enhancing pleasure in response to art.

The gallery is another 'sex show' location promoting and delighting in the sale of women's bodies (and plastic 'buckets' modelled on women's bodies) under the badge of an economic analysis of working conditions and worker's rights. The viewer of *Maintenancer* in the art gallery is invited to watch, imagine, intellectualise, theorise and enjoy the pleasurable sensation of 'the feeling of life': the self-affirmation of art appreciation. The artists and curators of *Maintenancer* likely distance themselves from Kantian aesthetics and might invoke instead political and theoretical references and direct connections to the world. However, the eroticised art buzz of such supposedly transgressive subject matter is possible because the representation is a long way from the real ties of prostitution, which are revealed by radical feminist analysis and the accounts of exited women (Moran 2013; Döring 2022).

In the 1800s, the doctor Alexandre Parent-Duchâtelet "described the whore (sic) as a 'woman-sewer' and the brothel as 'seminal drain'" (Showalter 1990: 193). *Maintenancer* jumbles objects and human beings and like Parent-Duchâtelet treats prostitution as a social inevitability and necessity. He referred to women as 'sewers'; *Maintenancer* casts them as 'care workers' but both deal with the same male sexual behaviours and body fluids.

Ekman argues that contemporary prostitution, by demanding emotional engagement—'care work,' 'therapy,' 'the girlfriend experience'—invades women's selfhood and blurs the prostitute relationship with other relationships: "the more 'human' prostitution appears, the more devastating it is for the seller, because she is required to sell her humanity, her Self" (Ekman 2013: 110).

Instead of a bucket, other 'objects' are supplied: sex dolls and sex robots, and not only the more readily available women's bodies but also women's selves.

The 'Talking Woman' and the Porn Voice

The encroachment upon women's selves can also be found in the sexual commodification of women's voices through audio technology. In the first days of commercial phonography, recordings of women's voices were unpopular. A letter writer to the American *Radio Broadcast* magazine in August 1924 reported:

> ... that the public refused to buy recordings of women talking. Manufacturers lost several thousands of dollars, he wrote, before they learned that the public will not pay money to listen to the talking record of a woman's voice. His interpretation was that the voice of a woman when she cannot be seen "is very undesirable and to many, both men and women, displeasing". (McKay 1988: 171)

This was despite the success of women as monologuists in vaudeville theatre (171) but might have correlated with a tendency, in the early years of the twentieth century, "for women on the burlesque stage to become increasingly mute spectacles" in the form of strip tease, or as the butt of sexual humour voiced by male performers (Smith 2004: 26). The shift to sexualise and silence the women was so pronounced that the term 'talking woman' was used for the unusual occasions upon which women performers spoke.

A 'talking woman' is central to *Silent George*,[5] a 'party disc' (undated, possibly 1950s) in which a female voice narrates a rape and expresses sexual arousal. Between the 1930s and 1950s adult 'party discs' or 'blue discs' were produced and sold 'under the counter.' These were spoken and vocal recordings, by male and female actors, of: sexual innuendo and jokes, often about female genitals; sexually ambiguous narratives; and sexually explicit scenarios involving women voicing arousal sounds. "The most frequent type of blue disc presents a woman who is seemingly unaware of the erotic nature of the dialogue in which she is involved" (Smith 2004: 29). In these a woman's non-sexual speech is reframed by the male performer as sexual; her voice is used against her and to sexualise her.

In *Silent George* a male voice jovially welcomes listeners to "a story about a man known to his many girlfriends as Silent George" and tells us we are in the bedroom of a young girl asleep at midnight. The remainder of the recording consists solely of the speech, pleading, shouting and arousal sounds of a woman. She speaks as if to Silent George, a man she apparently knows, who enters her room uninvited and starts to undress and assault her: *you shouldn't do it, no don't.... please you're tearing my new nighty... please don't put your hand down there*. Sometimes she bargains with him: if she lets him kiss her breast, will he go; if she lets him *play around for two minutes,* will he go. At one point she tells him he is making her excited. She asks what he *is going to do with that jar of Vaseline*, tells him *I am not going to uncross my legs*... says *you're forcing me*, threatens to *call for daddy,* says *you're hurting me,* says *now I am going to scream for daddy*. Immediately at the first shout of 'daddy' her voice slackens off and the repeated 'daddy oh daddy's become intimate moans, voice cracks and panting breath.

Silent George's story, a standard porn narrative—of the woman who is forced and who simultaneously 'really wants it'—is communicated entirely through the voice of the woman. Her disembodied voice voices the sexual

[5] The story line of *Silent George* appears to be an elaboration of a song written in 1950 by Henry Glover and Sally Nix and performed by Wynonie Harris and Lucky Millinder Other musicians have also made versions.

objectification of her body. She speaks, pleads and moans directly to the listener who is positioned as Silent George. She speaks his fantasy for him.

The film *When Harry Met Sally* (1989) placed the sound of a woman's orgasm into the centre of public space. The character Sally spontaneously performs 'an orgasm' in a very crowded restaurant. It is presented as an act of female empowerment. To 'prove' to her friend Harry that he cannot tell whether a woman is really having or pretending to have an orgasm, Sally pretends to have an orgasm sitting opposite him at the table. It is primarily a vocal performance. This infamous film scene placed the unambiguous sound of a woman's 'pornographic voice' into the public spaces of the restaurant in the film, the cinema auditorium and the mainstream television broadcast.

The sounds of pornography, however, in the form of moaning and panting women's voices were already public and widespread since the 1960s through popular music. Sound artists and writers John Corbett and Terri Kapsalis refer to these sounds as "aural sex." They note that by the 1990s digital sampling technologies made "female sex vocalisations (moans, shrieks, gasps, sighs) ... a staple of dance music" (Corbett and Kapsalis 1996: 183). Numerous pop songs that include "female orgasm sounds" could be heard, as background music, in public places such as shops, workplaces or on the radio. It was "sex in places you aren't allowed to see it" (185). "The explicit sex sounds used in popular music are clearly often a direct genre reference to mainstream cinematic and videographic pornography" (187).

Like many pro-porn arguments of the 1990s (Williams 1990; Bell 1991) Corbett and Kapsalis' discussion is framed in relation to a psychoanalytically influenced proposition about the disruptive potential of 'female pleasure.' The theory is that female pleasure has been psychologically repressed and socially prohibited and therefore that it is sexually radical and progressive to explicitly represent it. Corbett and Kapsalis propose that "the public airing of such graphic sex sounds" might cause "profound disturbance, a gentle threat to the stability of sensical representation .... unsettling to deep cultural architecture" (182–183). The language they use eroticises the argument itself. They embellish this with ambiguity (the truth/lie problem of the sound of female orgasm) and word play on female (vocal/sexual) ejaculations. They create a kind of theory-erotica to legitimise an increasing pornification of everyday culture.

Nonetheless, much of their commentary usefully reveals the realities of the pornography structure of these examples of pop music. When they refer to women's orgasm sounds as "infinitely repeatable and renewable resources" they position women's bodies as an exploitable resource, and they invoke the sexological trope of the multi-orgasmic sexually insatiable woman (Coveney et al. 1984: 97). They identify a "sadistic listening position" and the "enjoyment of the sound of a woman in pain" (191).[6] They observe that:

> In almost every example … female sexual vocalisations blurred the line that separates a representation of pain from a representation of pleasure, often sounding uncomfortably like screams of torture as much as outbursts of sexual pleasure. (191)

They also describe a "double sex/tech fetishization" (189) evident in product descriptions of non-musical LP and CD recordings made as audio pornography. Technologised sexual access is promised by innovative recording techniques and talented sound engineers: *high fidelity … a noteworthy landmark in recorded sound …most advanced electronic techniques … virtual audio engineering … virtual 3D audio… first virtual reality sex experience … Cyborgasm is so in-your-libido vivid … sounds so real you're not just hearing sex, you're having it* (188–189). It is the same double sex/tech fetishisation as is used in the marketing and mythologising of the sex robot.

Telephone Sex

In the 1980s and 1990s in America and Europe, a combination of new telecommunication and payment methods enabled a proliferation of telephone sex lines and dial up porn services. Women were encouraged to think of this as an opportunity to 'sell sex' safely without risk or violation.

One night in April 1984 a DJ on a freeform experimental radio station in New Jersey rang a telephone sex line live on air and broadcast the conversation.

[6] They also propose that there is a masochistic identification by the male listener with the woman-as-victim.

Frank Balesteri asks the woman, who uses the name Tabatha and whose speech he pays for, to incorporate particular details into her sexual performance. In line with his kitsch radio-artist persona he tells her to talk to him about Elvis Presley, to sing Elvis songs, to describe fellating Elvis, to lick ejaculate off Elvis' blue suede shoes. Uncertain, off guard and nervously half-laughing she does what he says and incorporates it all into her routine. Her routine is to use her voice to describe the sexual use of her body, to aggrandise the sexual ego of the buyer and to deliver to him his orgasm within the prepaid time slot. A buyer can say anything to her, ask anything of her.

The DJ casually exploits the fact that this woman's voice, time and attention are available to him and he uses this to make a public piece of experimental radio. He demonstrates that he can pay for her sexual availability and share her sexually with his listeners; he can use her voice as part of his creative work and inflate his status as a transgressive artist.

Decades later the recording of her voice continues to circulate as one of a 'best of' album of audio tracks online and transcribed and published in a book about avant-garde radio culture (Ballesteri (sic) 1993: 157–159).

Sheila Jeffreys points out that phone sex lines allow men to buy the opportunity to verbally harass women and that they "institutionalise the practice of 'obscene phone calls'" (Jeffreys 1997: 267), which are used by men to violate women. As a whole "the sex industry markets precisely the violence, the practices of subordination, that feminists seek to eliminate from the streets, workplaces and bedrooms" (267–268).

The structure of the 'obscene' or 'dirty' phone call is forced sexual intimacy in which a (usually anonymous) man says something sexual or makes sexualised vocal/breath sounds in response to the voice of the woman (or child) who answers the call. A phone speaker is held close to the ear, and this is exploited to create the visceral shock of intrusion. The caller's sexual gratification is achieved through his invasion of boundaries and his eroticisation of the woman's response be it fear, confusion, protest or silence. The sexual harassment extends beyond the immediate call. Obscene callers "shatter our security in our home" (Coveney et al. 1984: 18) and often use the terror technique of suggesting they are an acquaintance, leading women to wonder whom they can trust. The 'dirty' phone call is a signal and act of male social control of women.

Mobile phones, social media apps and conferencing platforms now enable a wider variety of 'dirty' calls. Men now 'flash' via communication technology—"intrude their penises on ... women's attention" to frighten and offend them (Jeffreys 1990: 123). Jeffreys notes that most twentieth-century sexologists considered flashing to be a "harmless male hobby." It is a 'hobby' which has increased exponentially among men using contemporary technology.

In '"It makes my skin crawl": sexual moaning rise in schools' feminist writer and activist Caitlin Roper reports on the current widespread incidence, in Australian and American schools, of another type of voice related sexual harassment: sexual moaning. Girls and female teachers report how male pupils make "sexual moaning, groaning and grunting noises" directly and up close into the ears of female pupils, in the classroom, on school buses, etc.

> I have also had female teachers approach me in tears, describing what it feels like to be greeted by these orgasm noises every time they enter the classroom. Sometimes the panting noises are played from porn videos through their devices. [...] I've dealt with male teen students ... They play the female fake orgasm noises - it causes great discomfort for female students, but the girls are learning to grin and bear it. (Roper 2022)

Roper describes how school authorities do not take action to prevent the behaviour. Instead, teachers explain that the authorities treat it as normal development for boys, thereby validating porn use, porn imitation, verbal assault and the sexual harassment and silencing of women and girls.

As Kappeler says, pornography viewing is also doing. Women's voices which have already been sexually objectified, bought and sold and masturbated to by men, are "sound grabbed" and used against women in everyday public social realms.

The men and boys who watch and play porn on their mobile phones in schools and workplaces and on public transport do so because they can. They are not doing this secretly; they want women to know that they can use porn wherever they choose. They aim to sexualise, intimidate and humiliate women by doing so. They make sure women can hear or see the porn, to show how very well established is the male sex right. Porn is used to silence women's voices. Men's use of porn is intended to teach women to 'grin and bear it.'

Pornography Scenarios: The Unwilling and the Willing Victim

Kappeler describes how pornography has two archetypal victim scenarios: one, the unwilling victim and two, the willing victim. "Scenario no. 1 is a reflection of the male culture's historical obsession with female virginity" (Kappeler 1986: 157). The plot of scenario one is that a woman is raped and resists and screams. In scenario two, instead of resisting and screaming, she enjoys the rape, and orgasms.

In the masculinist (anti-feminist) narrative of pornography, she can choose to scream (and 'be a victim') or choose instead to orgasm and become a collaborator in the pornographer's project. The plot of scenario two is that a woman is sexually 'fulfilled' by rape.

By choosing to orgasm, she 'escapes' prudery, repression and virginity, and proves that 'she wanted it.' In both scenarios woman is a victim-object of male-dominant sexual behaviours. The pornographic subtext of scenario two (or of the shift from scenario one to two) is that her body tells 'the truth' (that she wants to be raped) and overcomes 'the lie' of her voice (that she doesn't).

The scenarios mirror the stories of the Marquis de Sade's two characters Justine and Juliette. Justine is sadistically raped, brutalised and humiliated and screams. Juliette is sadistically raped, brutalised and humiliated and 'chooses' to orgasm and collaborate with the sadist.

In the scenario of the unwilling victim "… the hero pursues his pleasure with brutality, inflicts pain regardless. It is expected that the victim cry out" (137–138).

In the scenario of the willing victim:

> .. for his (the hero's) pleasure, for his sexuality, he must have the pleasure of the woman, the 'emancipated' woman. The liberated woman, the one liberated by Sade's imperative: you shall enjoy (your duty, your pleasure), do not withhold yourself. The hero wills his woman's lust. He is obsessed with her orgasms. (138)

He says: "she may have an orgasm, but 'I have a girl': I have her and I have her orgasm" (139).

These scenarios map not only onto the narratives of pornography (which are 'rape' and 'the woman wants to be raped'), but also onto the sexual revolution, contemporary western sex culture and the pre-occupations of a considerable amount of twentieth century/twenty-first century critical and literary theory.

Post-sexual revolution, the orgasm is a woman's obligation, a duty, just as sex was the duty of the pre-sexual revolution wife (Jeffreys 1990). The 'willing victim scenario' is embedded in so-called 'feminist porn,' BDSM, pro-kink culture and the argument that the problem with prostitution is stigmatising it as prostitution rather than sanitising it as 'sex work.' Similarly, waves of post-modernism, post-structuralist and psychoanalytic informed theory have fixated on "the pursuit of pleasure ... the play of desire" (137) and an eroticised conceptual 'femininity' (or 'otherness'). In theory-erotica philosophers trade in theory-orgasms.

Beatrice and the sex robot share archetypal love-and-sex 'victimhood.' They are both ideal, exquisite man-made artworks, a fictional character eulogised as sculpture and a manufactured figurine. One is a soul mate, one is custom-made-for-you; both are created in eras of exploding pornography technology. They both represent 'child-women,' 'bodies' without experiences, without life. One dies a 'romantic death,' one is a dead woman body object, animated. They both represent the abandonment which verifies male sovereignty and inflames 'the feeling of life.' Their voices both belong to men.

Beatrice would be an unwilling victim of rape but willingly consents to love-as-death. In death, she screams through the narrator's body. Beatrice becomes only screams in his body.

The sex robot is an idolisation of the willing victim: it has no feeling, no pain, no pleasure, but makes 'female' orgasm sounds for a man's gratification. An owner 'has a girl' and 'has her orgasms.'

In Plain Sight: Scream Collecting and the Library of Orgasms

Roland Barthes calls the scream a "sonorous fetish" and he notes "the value of a machine to isolate the scream and deliver it to the libertine as a delightful part of the victim's body" (Barthes 1976: 143). He refers to a piece of torture apparatus described by Sade: a "tubed helmet," worn by a woman being sexually abused, which "transmits her pain" to the listeners in an adjacent room "as if by radio" (144). The technology links the torturer to his libertine guests. Sade describes exactly how technologies will continue to be deployed, in the centuries to come, to exploit women's voices.[7] A tortured woman's screams are not her own.

Bijoux Indiscrets is on online sex shop aimed at women. It is named after Denis Diderot's eighteenth-century novel *Les Bijoux Indiscrets,* translated into English as *The Talking Jewels* or *The Indiscrete Toys*. The novel tells the story of a sultan given a magic ring which enables him to be invisible and to command any woman's vagina to speak and tell the truth about her sexual experience. The shop sells sex toys, creams and bondage straps, chains and restraints described as body jewellery. It is branded like a boutique or a lifestyle and education resource promoting emancipation, sexual well-being and female pleasure. It includes research and statistics, gathered from women and men, about sexual expectations and behaviour. The founders say: "Bijoux Indiscrets is made up of women prepared to push the boundaries of female sexuality much further than they've ever been pushed before." It is marketed to Kappeler's 'willing victim': the woman whose response to the vexation of contemporary sex culture is not to scream but to figure out how to have more orgasms.

It also hosts an "ORGASM LIBRARY OF REAL SOUNDS." Established in 2016, this is an archive of recordings of women's voice sounds. The website page reads:

> *Help us show the diversity of female pleasure by sharing our project. You can take part, too. Upload your orgasm.*
>
> *UPLOAD ORGASM - 100% REAL · TOTALLY ANONYMOUS* (Orgasm Library and Indiscrets 2022)

The library project claims to be in the interests of women and aligns with the values and intentions of the shop, such as narrowing "The Orgasm Gap." It invokes a mixture of sexology, self-help and exhibitionism. A woman is encouraged to think that 'uploading her orgasm' will contribute to a female sexual progress movement. She is encouraged to breach the boundary of her sexual privacy and to consider 'orgasm upload' as an uplifting sexual adventure.

There are more than two thousand uploaded recordings, searchable via porn-type hashtags. They primarily feature closely recorded women's

[7] Numerous films centre a woman's scream as the dramatic climax or truth teller.

arousal sounds, with a few variations.[8] Ultimately, it is a collection of audio pornography, by which women's 'uploaded orgasms' are made available to the international digital world.

The *Legal Notice* and *Terms and Conditions* are revealing. Those who upload recordings "cede all intellectual property rights in the orgasm audios" forever, anywhere. The website owner is granted "the right to the exploitation of the aforementioned orgasm audios" and shares the same rights—to transmit, publish and distribute—with anyone who downloads or copies the sounds.

The women's voices and breath in the Library of Orgasms are public property, over which they have no further claim. Presumably, the erotic conflation of toys, jewels, vaginas and indiscretion appealed as a brand name, but *Les Bijoux Indiscrets* is a story about a man's coercion, voyeurism and unrestricted access to women's sexual lives.[9] In plain sight, the sovereignty of male sex right is being sold to women as female sexual emancipation.

The Cult of Dissociation and Dummy-Women

The death and alienation celebrated by the decadent fin-de-siècle cult of syphilis and artifice are repeated in the cult of dissociation found in contemporary technology and sexuality. Two final examples demonstrate the lineage.

In 1884 J.-K. Huysmans published *A Rebours,* an influential decadent literary classic (titled in English *Against Nature*). The sexually jaded protagonist Des Esseintes, like contemporary transhumanists, is preoccupied by esoteric possibilities of exchange between artifice and reality. Des Esseintes sees a female ventriloquist perform in a café and this provokes "a whole crop of new ideas." "[He] lost no time in firing off a broadside of banknotes to subjugate the ventriloquist" (Huysmans [1884] 1959: 113). His "new ideas" are vocal and sexual performances which can be achieved using the ventriloquist's "monstrous speciality" (113). Des Esseintes first requires that, while she lies beside him, she ventriloquises

[8] There are some in which man and women are audible; a few are solo male voice sounds; there is one like a horror soundtrack with a woman's screaming, a 'diabolical' voice and crashing sounds.

[9] Not unlike the methods of sexological research and the justice system dealing with MVAWG.

two figurines, a marble sphinx and a terracotta chimera, with dialogue from Flaubert's *The Temptation of St Anthony*. Des Esseintes feels himself the direct addressee of the chimera and thrills at his communion with the words of Flaubert. The woman is used for the aesthetic meeting of men: through her disembodied voice the male writer and male audience commune.

Later, to sexually excite him, he demands that she ventriloquise "a husky, drunken voice" roaring at the door of the bedroom: "Open up, damn you! I know you've got a cully in there with you! But just you wait a minute, you slut, and you'll get what's coming to you!" (115).

Thrilled by the threat of intrusion Des Esseintes would "hurl himself upon the ventriloquist, whose voice went blustering on outside the room" (115). His sexual gratification is achieved by paying the ventriloquist for the use of her body while she verbally abuses herself with her dissociated voice.

Huysmans describes women with disgust, contempt and dismissiveness. For Des Esseintes, the ventriloquist is the most memorable, "her artifices made other women seem insipid" (115). Her voice was dislocated from her body, projected into monster objects, thrown outside the room and spoke exactly as he instructed. She has no name, and although she has the technical ventriloquial skill, he has the power of the male subject to make her say what he wants her to say. In a man's voice, she calls herself a slut and threatens herself. He makes her into his sexual dummy and he gets her to voice his speech—the same sexual misogyny and threats of violent punishment that are used against women in contemporary pornography.

In 2019, American comedian Whitney Cummings commissioned an AI enabled sex robot portrait of herself which she incorporated into her stand-up routine and talk-show interviews. She talks to the robot which responds with a mixture of pre-programmed and algorithmically generated material.

The sex robot is both 'transgressive' and 'humourously' underwhelming. Cummings' supposedly ironic critique of it instead lends layers of nuance and credibility to its cultural status. She makes a self-deprecating joke about how her boyfriend hates the doll because he says one of her is already too much. Then she recounts his discovery, in her absence, that the doll's breasts feel realistic: implying that when she was away her boyfriend might have used the 'hated' sex robot sexually.

Both she and the doll wear practical clothing, jeans, long sleeved shirts and hoodies. However, we know the sex robot is a sex robot, an object man-made for sex. It exists because of financial investment in woman-objectifying sexual behaviour. Even when not sexy it is sex, just as women are seen as synonymous with sex, irrespective of whether they are considered sexy.

Having given a lot of money ($100,000 she repeatedly tells us) to the sex industry, Cummings gets to play at being in on the joke, being a collaborator, a libertine, as Kappeler would say, but this in no way subverts the joke; it just makes the 'joke' on women last longer (Netflix is a Joke 2019; Rich Eisen Show 2019).

A woman bringing the robot into public space is strategic for the industry. Public, visible female ownership of dolls popularises and normalises the doll and the doll industry and invariably it plays on the idea that the dolls might be like the women, or the women like dolls.

Cummings is a stand-up comedian, and her voice is her stock in trade. She says herself the sex robot is a gimmick. Question: how does a woman get attention? how does she get her voice heard? Answer: by commissioning a speaking sex-object likeness of herself and performing with it. Whitney Cummings' voice appears to have cultural self-determination but it is used just as Tabatha's is by the radio DJ.

Body and Breath

In the contemporary cult of dissociation, transhumanist language for the human body includes terms such as 'meat sack' and 'wetware.' These suggest conceptualisations of the body very similar to the 'woman-sewer': the dehumanised body as a convenient, expendable container, like a peep-show bucket. While transhumanists diminish the biological body, materials technologists aim to make sex doll bodies skin-like, meaty and wet. The sex robot is sold as a triumph of art and science and the apogee of combined sex and tech fetishism. More economically, Realdolls' social media promotes masturbation devices shaped like half a woman's face with the line: "Sometimes it's nice to deal with a mouth that doesn't talk back."[10]

Beatrice's voice and life are taken away in her breath. Today 'breath play,' the eroticisation of restricted breathing, is presented to young women by mainstream sex educators and liberal feminists, as a normal

[10] https://twitter.com/AbyssCreations/status/1397266653541330947.

sexual activity. The seventies porn fixation on 'deep throat' has flourished and been joined by other sexual practices which attack women's throats. Sexual gagging, strangulation and choking all limit a woman's voice and breath and threaten to take her life. There is a new market for breast binders, which damage young women's ribs, posture and lung capacity. The breast binder is a twenty-first century variation in the fashion/fetish history of corsets for women. Jeffreys comments on contemporary corset use: "Restricting women's ability to breathe seems to be an important aim of heterosexual male corset fetishists" (2005: 93).

Beatrice's narrator kills himself to abolish her voice. He is the abuser who kills a partner, or wife and children, and then himself in the domestic murder-suicide. He epitomises the ultimate life and self-destroying logic of extreme misogyny. The 'intimate' annihilator is the narrator of his sovereign world. He sees women as either possession or part of himself. As narrator, he controls his story and the lives of his possessions.

'Lovefaking'

One of the most powerful pieces of knowledge that domestic violence organisations offer to women is that abusers tactically alienate women from their own instincts and intuition, their inner voices.

> You are unaware that the very basis for your decision making, your facility for instinctive reaction is unavailable to you [...] You lose the ability to make quick decisions and your animal brain becomes so quiet that your only response to danger is to freeze. (Hennessy 2018: 43)

Likewise, psychoanalysis peddles the story that women cannot know themselves, what has happened to them nor what they really want. It is hard to overestimate how deeply the 'stories' of psychoanalysis have entered norms of thought, our 'folktales'[11] about sex and selfhood. They shape the story of sexual liberation into the 'healthy embrace' of new sexual practices ('sex-positivity') in defiance of the 'terrible threat' of repression. They explain that if you react with resistance, it is the very proof of the power of your repressed desire. You cannot say no and mean it. Your unwillingness indicates your willingness. Psychoanalytic

[11] Kappeler uses the term 'cultural folktale' to describe the patriarchal story that keeps being repeated.

truisms—that women's repression and societal conservatism impede full sexual self-actualisation—lead to individual and social tolerance of obviously dangerous and damaging behaviour. The credo of the Freudian unconscious, in popular 'wisdom,' displaces the voices of both instinctual gut-feeling and feminist political consciousness: it attacks women's inner and public voices.

Freud, sexologists, artists, writers, transhumanists[12] and the sex industry ventriloquise women's voices. The sex-robot-peep-show-bucket is voiced by technologists, buyers and pornographers. It is emblematic of how women's voices are turned against them and an example of how technologies—mechanical, optical, digital, medical and surgical—are used in the service of increasingly efficient and widespread modes of pornography, prostitution and clinical control of women's sexual, reproductive bodies.

Women's voices are stolen, silenced, borrowed, respoken, replaced and coerced. Men give and receive and use women and their voices between themselves. Schwob writes *Beatrice* for his male reader; Huysmans' ventriloquist connects Des Esseintes and Flaubert; Tabatha is shared between the telephone-sex-pimp, the DJ, and his listeners; Emma Eckstein was passed between Freud and Fliess to be mutilated, manipulated and co-opted into the raison d'etre of psychoanalysis.

'Decadent' men present women as the willing victims of voice theft and use their voices for sexual pleasure. It is a hellish hallucinatory reality when a 'speaking' peep-show bucket is presented as a love partner. Hillel Schwartz describes the sex doll as a "klein bottle that neither gives nor receives" and interaction with it as "lovefaking detached from the personal rhythms" of human embodied, interpersonal life (1996: 129).

Kappeler wonders why we rarely ask, not what the pornographer is representing, but what the pornographer is really saying. We might answer. He says: I am. He says: I have freedom of speech.

Postscript

On 21 July 2022, the BBC posted a short video article in response to the announcement that "Medical students in England will have to

[12] Bina48 is a speaking robotic head modelled on Bina Aspen, wife of transhumanist Martine Rothblatt. The head is designed not only to speak like Bina Aspen but as a repository for her consciousness in the future.

pass mandatory training to diagnose and treat female health conditions under plans to improve women's health" (*BBC News* 2022). Three women discuss the difficulties faced in getting healthcare for gynaecological conditions such as endometriosis. In summary, one says: "Let's face it. You've either got a vagina or a voice. You don't seem to have both, do you?".

References

Ballesteri (sic), Frank. 1993. The vanilla bean talks to a phone hooker about Elvis live on WFMU. In *Radiotext(e)*, ed. Neil Strauss. New York: Semiotext(e).

Barry, Kathleen. 1995. *The prostitution of sexuality: The global exploitation of women*. New York and London: New York University Press.

Barthes, Roland. [1971] 1976. *Sade/Fourier/Loyola*, trans. Richard Miller. New York: Hill and Wang.

BBC News. 2022. Women's health: "I had periods that lasted for months". [online], July 21. https://www.bbc.co.uk/news/av/health-62229951. Accessed 22 Jul 2022.

Beauvoir, Simone de. [1949] 1953. *The second sex*, trans. H.M. Parshley. London: Jonathan Cape.

Bell, Shannon. 1991. Female ejaculations. In *The hysterical male: New feminist theory*, eds. Arthur Kroker and Marilouise Kroker. London: Macmillan.

Corbett, John and Terry Kapsalis. 1996. Aural sex: The female orgasm in popular sound. In *Microgroove: Forays into other music*, ed. John Corbett. Durham and London: Duke University Press.

Coveney, Lal, Margaret Jackson, Sheila Jeffreys, Leslie Kay, and Pat Maloney. 1984. *The sexuality papers: Male sexuality and the social control of women*. London: Hutchinson.

Döring, Mia. 2022. *Any girl: A memoir of sexual exploitation and recovery*. Ireland: Hatchette.

Ekman, Kajsa Ekis. 2013. *Being and being bought: Prostitution, surrogacy and the split self*. Melbourne: Spinifex Press.

Green, John A. 1965. Marcel Schwob and the talking machine—A tale a la Poe via Thomas Edison. *BYU Studies Quarterly* 6 (1): 41–47.

Hall, James. [1974] 1983. *Hall's dictionary of subjects and symbols in art*, 2nd ed. London: John Murray.

Hennessy, Don. 2018. *Steps to freedom: Escaping intimate control*. Dublin: Liberties Press.

Herman, Judith Lewis. [1992] 1994. *Trauma and recovery: From domestic abuse to political terror*. London: Pandora.

Huysmans, Joris-Karl. [1884] 1959. *Against nature*, trans. Robert Baldick. Middlesex: Penguin.

Jeffreys, Sheila. 1985. *The spinster and her enemies: Feminism and sexuality 1880–1930*. London: Pandora.

Jeffreys, Sheila. 1987. *The sexuality debates*. London: Routledge & Kegan Paul.

Jeffreys, Sheila. 1990. *Anticlimax: A feminist perspective on the sexual revolution*. London: The Women's Press.

Jeffreys, Sheila. 1997. *The idea of prostitution*. Melbourne: Spinifex.

Jeffreys, Sheila. 2005. *Beauty and misogyny: Harmful cultural practices in the west*. London: Routledge.

Kappeler, Suzanne. 1986. *The pornography of representation*. Cambridge: Polity.

Kilby, Benison. 2020. Sex work, care work and art work in Sidsel Meinecke Hansen and Therese Henningsen's Maintenancer (2018). *un Magazine* 14 (1): 114–119. https://rodeo-gallery.com/documents/93/Sex_Work_Care_Work_and_Art_Work_Benison_Kilby_un_Magazine_2020.pdf. Accessed 25 Jun 2022.

Masson, Jeffrey. 1992. *The assault on truth: Freud and child sexual abuse*. London: Fontana.

McKay, Anne. 1988. Speaking up: Voice amplification and women's struggle for public expression. In *Technology and women's voices*, ed. Cheris Kramarae. London: Routledge and Kegan Paul.

Millet, Kate. [1969] 1977. *Sexual politics*. London: Virago.

Moran, Rachel. 2013. Paid for: My journey through prostitution. Dublin: Gill & Macmillan.

Museum of Modern Art. 1967. *History of the mutoscope*. (Archived press release). https://www.moma.org/momaorg/shared/pdfs/docs/press_archives/3932/releases/MOMA_1967_July-December_0014_82.pdf. Accessed 25 Jun 2022.

Needham, Gerald. 1972. Manet, 'Olympia' and pornographic photography. In *Woman as sex object: Studies in erotic art*, ed. Thomas B. Ness and Linda Nochlin, 1730–1970. New York: Newsweek.

Netflix is a Joke. 2019. *Whitney cummings sex robot unveiled 7/8*. https://www.youtube.com/watch?v=QPPkK9zexvY. Accessed 22 Jul 2022.

Nyholm, Sven and Lily Eva Frank. 2017. From sex robots to love robots: Is mutual love with a robot possible? In *Robot sex: Social and ethical implications*, eds. John Danahar and Neil McArthur. New York: MIT Press.

Orgasm Library and Bijoux Indiscrets. 2022. https://shopeu.bijouxindiscrets.com/pages/our-philosophy https://orgasmsoundlibrary.com/#gallery (websites: online shop and 'library'). Accessed 2 Jul 2022.

Pankhurst, Christabel. [1913] 1987. The Great Scourge and how to end it. In *The sexuality debates*, ed. Sheila Jeffreys. London: Routledge and Kegan Paul.

Rexer, Raisa. 2021. *The fallen veil: A literary and cultural history of the photographic nude in nineteenth century France*. Pennsylvania University of Pennsylvania Press.

Rich Eisen Show. 2019. Whitney cummings (and her robot clone) Talk new Netflix special. Aug 26. https://www.youtube.com/watch?v=Xi0HsmnDafI. Accessed 22 Jul 2022.

Roper, Caitlin. 2022. "It makes my skin crawl": Sexual moaning rise in schools. *Collective Shout*, Jun 9. https://www.collectiveshout.org/sexual_moaning_rise_in_schools. Accessed 1 Jul 2022.

Schwartz, Hillel. 1996. *The culture of the copy: Striking likenesses unreasonable facsimiles*. New York: Zone.

Schwob, Marcel. [1891] 1992. Beatrice. In *The second Dedalus book of decadence: The black feast*, ed. Brian Stableford. Cambridge: Dedalus, 28[illegible]–287.

Showalter, Elaine. 1990. *Sexual anarchy: Gender and culture at the Fin de Siècle*. New York: Penguin.

Smith, Jacob. 2004. 'Filling the Embarrassment of Silence: Erotic Performance on "Blue Discs"'. *Film Quarterly*. Winter. 58 (2). 26–35

Spender, Dale. 1983. *Feminist theories*. London: The Women's Press.

Unknown artist. n.d. *Silent George*. (LP record) US: Party Time Records/Heirloom. https://www.youtube.com/watch?v=aDB_lyq6IDU and https://www.youtube.com/watch?v=3icCCjyRyYM. Accessed 29 Jun 2022.

Williams, Linda. 1990. *Hard core: Power, pleasure and the "Frenzy of the visible."* London: Pandora.

CHAPTER 9

The End of Sex Robots: Porn Robots and Representational Technologies of Women and Girls

Kathleen Richardson

In this chapter, I want to argue for the end of sex robots. I will problematise the use of the term "sex" that is associated with them and push back against attempts to dilute the meanings of sex to include *anything*. Initially, I considered a better description for these objects to be *masturbation robots*, but they function as a form of pornography, making them *porn robots: pornographic representations of women and girls*. As I will argue, it is a mistake to attribute sex to them as objects and the arguments I set out in this chapter also apply to porn dolls. Porn doll producers regularly pass off their objects as robots in their social media,

K. Richardson (✉)
Faculty of Computing, Engineering, and Media, De Montfort University, Leicester, UK
e-mail: kathleen.richardson@dmu.ac.uk

© The Author(s), under exclusive license to Springer Nature Switzerland AG 2022

K. Richardson and C. Odlind (eds.), *Man-Made Women*, Social and Cultural Studies of Robots and AI,
https://doi.org/10.1007/978-3-031-19381-1_9

advertising and marketing materials, even when there are no computational or engineering *special effects* to elaborate *more* porn in them (see MacWilliam's chapter on voice in this volume as an example of this). It is also porn doll companies that are at the forefront of producing a commercial market-ready version of a porn robot (see Realdoll.com).

In order to make sense of why these objects are porn, this chapter will provide a critical examination of what porn is, exploring how it functions as a cultural tool in reinforcing, and maintaining the structure of sex inequality of male domination and female subordination (Dworkin 1988, 1989). Since the 'porn wars' (1960s–1980s) porn has legal protection and exists as the 'free speech' of, arguably, men (MacKinnon 1993; Dworkin and MacKinnon 1988), and I will consider how decriminalisation of it has reshaped culture and society in detrimental ways by reconsidering marginalised feminist literature that warned of its harms.

Men and boys (a third of young people have seen porn by the age of 12[1]) are impacted in various different ways by porn and exposure to it is shown to impair their capacities to make mutual, empathetic, intimate, and loving sexual attachments with others (Dines 2010). Therefore, I will also argue that the abolition of porn is an ethically desirable goal for both sexes.

As founder of the Campaign Against Porn Robots (formerly the Campaign Against Sex Robots) I write from an experience of feeling like David tasked with the challenge of taking on Goliath—as many of the 'ethical' (for example invoking Kant and Aristotle) arguments to justify 'sex' robots mimic the defence of the porn and prostitution industries (Gordon and Nyholm 2022; Jecker 2021; Peeters and Haselager 2021). To what extent is the overwhelming push to protect porn robots from regulation a form of 'ethics washing'? This is an activity where those engaged in producing ethical narratives disguise or fail to disclose their underlying agenda (McNally 2015). In recent years, critics have taken issue with ethics used in the service of rationalising harmful technology agendas (Yeung et al. 2019), though little scrutiny has taken place on how a predominantly male tech community has used 'ethics' as a political cloak to justify defending the expansion of pornography in technological areas. As a case in point, this chapter is inspired by radical feminist ethics that prioritises the rights of women and girls to be free from sexual objectification, exploitation and violence in whatever forms they take. The very fact that robots in the form of women are not merely called

[1] See Culture Reframed: Building Resilience and Resistance to Hypersexualised media and porn. Culturereframed.com.

'robot women' but *sex robots* is a case in point. They are mimicking the role of women as sexualised objects: reduced to penetrable parts, dollified appearance and presented as pornographic speech. Porn is estimated to value over $90 billion dollars globally (Bindel 2019a), demonstrating its importance in the global capitalist economy and the reason for its continued existence, even if it harms relationships between people. In this chapter I prioritise a study of Representational Technologies of the Human (RTH) of women and girls.

Representational Technologies of the Human

Representational Technologies of the Human (RTH) is a concept and phrase I have developed as a result of studying robots and AI over the last twenty years (Richardson 2015, 2018). RTH is a mimesis-inspired term that clusters together those *propertied technological objects* that are derived from mimicry, imitation and representation of human existence and experience (Benjamin 1935 [1969]; Auerbach 1953; Taussig 1993). In accordance with its mimetic features, the concept encompasses a range of *made-up* artefacts and includes texts, dramatisations, art, porn robots, AI, anthropomorphic robots, digital avatars, deepfakes and other technologies where the referent is human. The relevance of mimesis to this field is crucial to its origin and continued form: 'The essence of the computer is the manipulation of symbols' (1963: 392) wrote Paul Armer in the classic text *Computers and Thought*. RTH are differentiated along many axes, one significant one is according to whether it is women and girls, or men and boys that are mimicked in the objects—with sex-role inequality *represented* in the making, presentation and use of them (Richardson 2016; Sparrow 2017).

RTH in the form of women and girls presents a new horizon for late capitalist accumulation built on female depersonalisation and dehumanisation.

Mainstream (arguably the orthodoxy today) academic theorising flattens distinctions between representations, property and humans (Haraway 2003; Latour 2012) and thus critical perspectives such as my own that affirm differences between humans and property are out on the margins (Lerner 1986). The margins are the places of radical resistance to *property relations*. Feminist author and social activist bell hooks recognised that ideas picked up by powerful institutions maintained power structures, she knew this when she discussed how the version of liberal feminism that came to dominate the academy and society was one that fitted with 'liberal individualism' and posed little threat to male power (2000: 9).

A mimesis lens regards these objects as *representations*, the origin of which comes from relations *between* humans, whereas academics in the field, Big Tech evangelists, and porn robot developers try to persuade us they can turn *made-up* property forms into beings or as good enough approximations thereby creating symmetries. These experts claim that words in a play that spawned the robot (R.U.R), are turned into 'real' things when they work on them, but they only produce *more fiction*. Therefore, my definition of a robot is an image (text, picture, etching, number) on a 2D or 3D surface. How did we get to the stage where images on surfaces and dramatised with computational engineering are seen as human substitutes? While the economic drive of RTH is to flatten distinctions between people and property (opening up new horizons of exploitation), this requires the *human imagination* (like a reader of a novel) to believe the made-up stories where the ultimate aim is a realisation of the merging—humans and property. Thus, unlike other technological pursuits (e.g., bridge building) a *surplus of fiction* is required for RTH ontology, purpose and rationale. David Levy provides a good account of this surplus: "imagine a world in which robots are just like us (almost)…in which the boundary between our perceptions of robots and our perceptions of fellow humans has become so blurred that most of us treat robots as though they are mental, social, and moral beings" (2007: 303)—he predicts 2050 as the year this will happen! Representational technologists are forever using this surplus of fiction as a means of elevating the status of their work. They know how to tell a ripping yarn and have had considerable experience in creating technological mythologies. In 1967, AI founder Marvin Minsky said "within a generation the problem of creating 'artificial intelligence' will be substantially solved" (2: sec.1.0).

Porn robots provide evidence of this *surplus of fiction* in operation characterised through their mimetic dispersion. Hence television dramas such as *Westworld* (2016) or *Humans* (2015) or films such as *The Stepford Wives* (1975) and *Blade Runner* (1982) feature actresses playing in the role of female robot characters but they are reduced to 'sex robots' due to how academics use these made-up stories *as evidence* alongside their own lofty ethical theories. RTH that mimic men such as The Terminator character from the eponymous film (played by Arnold Schwarzenegger who is pretending to be a robot) is not a pornographic representation of man (nor are the vast majority of RTH modelled on men and boys) but many of them underpinned by mimicry of the female sex *are* pornographic, and thus it is the abuse of women and girls that is foregrounded.

A critical study of pornography is required to contextualise the development of these objects.

Porn Robots 'Writing About Whores' Writing About the Abused

The term pornography is derived from two words—'*graphos*' (writing, etching, drawing) and '*pórnē*' (whore/prostitute) translated as 'writing about whores' (Dworkin 1989: 199). Radical feminist Andrea Dworkin writes "the *pórnē* was the cheapest (in the literal sense), least regarded, least protected of all women. She was, simply and clearly and absolutely, a sexual slave" (Dworkin 1989: 200). A *pórnē* is an inferior class and sex role *created* by a powerful male elite and refers to a prostituted slave in ancient Greece [twelfth to nineteenth centuries BC]. The slavers—males of the citizen class—created spaces called *brothels* where they could access and use their sex slaves. The '*pórnē*' then and now, is the explicit tie between property relations under slavery and capitalism and the operationalisation of sex inequality (male domination and female subordination). Another, and more accurate way of describing what porn is, becomes: *writing about the abused, writing about the sexually violated, writing about the prostituted*. Pornography by definition is not about mutual sexual intercourse, it does not arise from what is *between* its participants, but permits the use of (primarily) women and girls as sexual property by (primarily) men. In the temporal scheme, the abuse always comes *before* or *during* (with film and photography) the writing about it—its projection on surfaces. The *graphos* are representations of what goes on in porn. What goes on in it? The sexual abuse of the *pórnē* by a male with more economic and political power. How is this sexual abuse represented? It is expressed by being captured on the surfaces of material artefacts. Porn is transmitted through intermediary surfaces, and thus variations between texts, dramatisations, art, photography, film or animations is *formatic*—they are modes of writing. The writing occurs in various materials (plastic, metal, paper) allowing its expression on divergent surfaces.

In ancient Greece, the citizen male had the power to decide what 'sex' was, and what others were—labelling them 'slaves' and abusing women and children. The porn of that period was depicted on pottery (Keuls 1993) and a study of it can show us the links between the abuse and the writing of it on artefacts. The images on Greek pottery were examples of the 'sex' that took place at the behest of the citizen males who controlled the politics, culture and ideas of what occurred and explanations for it (see Aristotle's *the Politics*). The 'slaves' (an elitist categorisation to justify exploitation), wives or children were not volunteers in this process—but coerced, ruled and abused. Aristotelian ethics was also *the Politics* of the time, created to naturalise and justify the abuse.

In her classic book *The Reign of the Phallus*, Eva Keuls gives examples of the pornography in question. The pottery depicts various images such as 'molestation of women and men sexually harassing women' (image 213: 241), 'man negotiating the price of sex with a boy' (image 266: 297), 'mature man with responsive child lover' (image 254: 284), 'men molesting a naked hetaera' (image 155: 177), 'man threatening disobedient prostitute with a stick' (image 169: 185), 'man and older prostitutes' (image 167: 185), 'man penetrating doe' (image 161: 179) and all the other ways citizen males wrote about the abuse on the pottery and art they produced. At times they embellished the imagery with classical tales such as Zeus pursuing Ganymede (image 200: 225) or satyrs molesting women (image 214: 241).

Porn is produced when there is someone to abuse—someone who (predominantly) men can debase, dehumanise and strip of their dignity before *writing* (depicting, expressing) about it. When mediated in RTH the abuse occurs, this is then written about, and users get to engage with it. With a porn doll or robot they get an approximated version of the abused to elaborate it further.

Pornography's production is a mimetic process. It was once produced manually (writing, craft, art, pottery) and/or iterated through mechanical (photography and film) processes. Cultural historian Walter Benjamin describes the process of the technological development of representation in his essay *The Work of Art in the Age of Mechanical Reproduction* (1969 [1935]). I want to add computational (digital)to his list of processes. With the development of mimetic technologies, these graphic displays of subordination can be captured on different kinds of intermediary surfaces (pottery, films, computer chips). The development of a global plastic industry—a polymeric material that can be modelled in moulds—presents opportunities to make further kinds of pornography. Commentators are keen to note the pornography in robots is another version of its form in dolls and that while the surface may have changed, (the use of dolls was given to Dutch sailors in the seventeenth century (Smith 2013)), the content of the debasement remains the same, they are surfaces that present new opportunities to *write about the abused*—produced in contexts of masculine biases (Hicks 2017), sex slavery (Richardson 2016) and racism (Benjamin 2019).

Contradictory arguments can be used in support of porn robots.

On the one hand, advocates of porn robots and their infantile versions—child sex abuse dolls—propose these are symbolic and imaginary objects, and to regulate them is to rail against fantasy, to attempt to control the imagination and to police the inner lives of men who want to express themselves by using representational intermediaries and approximated versions (Harper and Lievesley 2020), while on the other hand, people and RTH are merged together. In terms of the former, they argue that this split-off world needs to be free from regulation, even supported to flourish as a cultural realm that can act as a depository for harmful (predominantly) male behaviours: rape, child sex abuse, torture—these are just some of the efficacious options this split-off world supposedly will help to support.

I am going to look at an example of how the ethical politicisation of 'fantasy' is mobilised to justify the existence of these pornographic RTH using the example of child sex abuse dolls. The child sex abuse doll is a surface for the writing of the abuse of children.

Prostasia is a US-based campaign organisation that describes itself as 'a new kind of child protection organization,' which values kink awareness, internet freedom and sexting law reform (https://prostasia.org/about/), they write:

> …leading scientists sincerely believe that the availability of sex dolls or robots may also prevent some real children from coming to harm'. (see Prostasia Foundation, n.d.)

Prostasia is currently running a funding campaign called 'Exploring the use of fictional and fantasy sexual outlets' and has raised 75 per cent of an $80,000 dollars target as of August 2022 (Prostasia Research Fund 2022).

In societies with child protection and an age of consent, children are off-limits to adults for their sexual use—this is termed *child sexual abuse*. A clear separation between adults and children recognises the multitude of ways in which children can be harmed by adult sexual predation: it can be carried out directly through contact, but also indirectly through the use of child abuse imagery (thereby also contributing to demand for more abuse imagery). Also child rights campaigners recognise that child abuse material can be expressed in other fictional ways (through animations or stories featuring child abuse). The legal demand to decriminalise (in countries where illegal) and extend the use of representational child abuse imagery should be seen as connected with the rise of the problematic term 'minor-attracted persons' (MAP).

What is a MAP? The 'minor' in MAP is a child under of the age of legal consent. The 'attracted person' is the adult who derives a form of arousal from the sexual abuse of a child. MAPs differentiate themselves by claiming they are not acting out the compulsion to abuse or rape a 'real child', only approximations of them. A worrying development is that MAPs claim this is their 'identity'. For adults to identify with practices that involve the abuse of children implies these are not lone individuals, but groups who discuss these issues and are working with academics who use ethics to do their politics of persuasion by arguing they present no harm to children. In fact, it is impossible to have a 'sexual identity' without consideration of what it entails. Sociologists and psychologists recognise the complex processes involved in sexual-identity formation, which requires experience, rehearsal, imagination, exploration and links 'human striving' (agency) and 'social organisation' (culture) (Côté and Levine 2014). To get to the stage where someone identifies as a 'MAP' raises serious child safeguarding issues. There is no space in this article to explore all the connotations of this debate (see Roper this volume), suffice to say for the purposes here, the use of the phrase MAP does two things. On the one hand, it broadens the definition of 'sex' to include children (Appel 2022). On the other, it creates a community of buyers and users for child sex abuse approximations (Harper and Lievesley 2020; Appel 2022). There is nothing intrinsically and necessarily harmful about representations of children—but it depends what underpins them—a doll in the form of a child with penetrable orifices offers adults an approximated experience of the sexual abuse of a child.

Attempts to split-off links between child sexual abuse in *vita reali* on the one hand and its representations on the other are afoot and present a serious threat to the safety, rights, dignity and protection legislation of children and their ability to live free from violence, and sexual objectification. The following is typical of how an argument for their use is presented:

> Rather than dismiss these technologies outright, our society has an ethical duty to explore their efficacy in carefully monitored interventions and to encourage them if they do in fact protect our most vulnerable from abuse and exploitation. (Appel 2022)

The examples above *separate* the abuse of children and its representations and can be exemplified by the 'ethics' argument *it's not a real child* (Appel 2022; Harper and Lievesley 2020). While I have given this example to illustrate the way in which ethicists separate abuse from representations, required when extending and protecting pornography, other fashionable academic philosophies collapse ontological distinctions between different phenomena.

For example, Donna Haraway's infamous *cyborg* figure presents an analytical framework to show how science and technology dissolve the boundaries between fact and fiction, writing 'This is a struggle over life and death, but the boundary between science fiction and social reality is an optical illusion'. Borrowing the ethos of this framework, one academic refers to the imagination as the giver of 'life' to objects writing of the "fantasy of intimacy with a 'brought to life-partner'—via doll personalization, modifications, online textual narratives and shared photographic poses" (Karaian 2022), implying the user of them ultimately determines their ontological status.

In another example exploring the experience of Davecat and his doll he has named Sidore, Nyholm and Frank (2017) write "The case of Davecat and Sidore suggest that people can experience themselves as falling in love with the sorts of synthetic sex dolls they might have sex with. So we can expect that some people will 'fall in love' with more advanced sex robots with which they have sex." They add a caveat to this: "Even if Davecat is very attached to his doll and thinks of himself as in love with 'her,' this is entirely one-sided…dolls cannot love anybody… *What if we fast-forward to a future that might not be too far off,* in which sex robots have become endowed with highly impressive forms of artificial intelligence?" (219 my emphasis). Leaving aside for the moment the surplus of fiction that is supplied by academics who support their development, if porn robots are both separate and merged at the same time, what kinds of objects are they?

Radical feminist analysis offers another way: if there were no sexual abuse, prositution or violation, there would be nothing to write about. The debasement of women is required for porn to exist at all (Dworkin 1988, 1989;Dworkin and MacKinnon 1988;Jeffreys 2005; Long 2012;Dines 2010; Bindel 2019b). These intermediary RTH and approximated versions are split-off from the *immediacy of the moment* in which the abuse takes place, but they are derived and exist because of the acts and are used as tokens of the abuse, repeated ad infinitum. Mack-innon extended this argument further and recognised porn provided a behavioural guide to men so that the women in porn might be viewed in 'two dimensions' but men who viewed it could act it out on women in their lives "their own three-dimensional bodies" (1993: 17). The widespread dissemination of porn has produced new forms of depersonalisation of women, including, arguably a collapse in differentiation between a woman as human, and woman as porn. The desire to delink porn from its representations in material objects (even if a film or image or sound recording depicting events) is central to the legal arguments that protected porn as 'free speech.' So actions (a man ramming his penis into a vagina) and recorded on film was classed as an 'idea' and an expression of man ramming his penis into a vagina (ibid.: 24).

The Truth About Porn and the Foresight of Pro-Women Feminists: Dworkin and MacKinnon

While porn has its antecedents in slave cultures pivoted on sex, class inequality and the abuse of children, the ubiquity of porn today is a relatively recent phenomenon and, in part, can be traced to the 'porn wars' in the 1960s through to the 1980s. In preparing this section, I had to confront the issue of a surplus of information on studies that demonstrate the harms of the porn industry which I cannot cover in its entirety. I have subsequently isolated what I consider to be some key issues related to the rise of techno-porn culture, of which porn robots are an exemplar. I want to return to the work of radical feminists such as Andrea Dworkin and Catharine MacKinnon and reconsider their work in light of the onslaught of pornographic RTH today.

In the 1950s, magazines like Playboy were a primary means in which pornography was disseminated, but by the 1960s, the pornographers had tapped into the 'free love' movement and rebranded porn as 'sexual liberation' to a new hippy generation (Jeffreys 1990). The male demand for porn increased and the technologised mass media infrastructure was utilised for its wider dissemination (Dines 2010). The demand for more porn, more availability and more freedom to view the kinds of activities carried out at one time by citizens of a slave class was now within reach of ordinary men. Porn was regulated and prohibited by obscenity laws and the push to make it legal began in earnest, resulting in the legal framework that has shaped the last five decades.

Porn arose in slave cultures where debasement of other humans was a fact of life by citizen males. Modern pornographers in the 1960s–1980s and their allies (Kaminer 2000: viii; Strossen 2000) had to create new justifications for the perpetuation of porn in so-called 'free and equal' societies marking it off as 'ideas' and 'expression' and protected as a First Amendment right to speech. In 1964, Justice Potter Stewart of the American Supreme Court is famously known to have said of pornography "I know it when I see it" (cited in MacKinnon 1993). This opaque statement has been referenced to argue that porn is—undefinable, a matter of taste (personal preference). Feminist legal theorist Catharine MacKinnon argued it also reflected a statement about men's power, of his viewpoint in knowing what it is. The legal framing of porn in this period has shaped the cultural narrative of it, as representations of sexual practices, that are in the "eye of the beholder," protected by an all-male Supreme Court (MacKinnon 1993).

MacKinnon and Dworkin tried to institute a civil rights ordinance on pornography that would give those in it who could prove to be harmed by it, the right to sue the pornographers—this was considered too much of a concession for the pornography industry which railed against it (even when the ordinances were legally voted for and passed in Minneapolis in 1983 and later in Indianapolis). The Minneapolis Ordinance stated:

> Pornography is sex discrimination. It exists in Minneapolis posing a substantial threat to the health, safety, peace, welfare, and equality of citizens in the community. Existing state and federal laws are inadequate to solve these problems in Minneapolis. (1998: 31)

As I explained earlier, pornography is ***writing about the abused***, a meaning which formed the basis for a legal definition developed by Dworkin and MacKinnon who needed to explain what happened in porn as precisely as possible, as laws need clear and concise meanings so they can be interpreted in disputes. Below is part of their extensive legal definition:

> ... the graphic sexually explicit subordination of women through pictures and/or words that also includes one or more of the following: (i) women are presented dehumanized as sexual objects, things or commodities.... (Dworkin and MacKinnon 1988: 36)

In order to write about the abuse, one has to refer to it. While pornographers write about the abuse they have done (do it and represent it, and share it with others to do it and share it) pro-women feminists write about the abuse for different reasons—to explicitly show the links between the writing, what underpins it, and how it is used as an instrument of power in every interpersonal relationship where porn is a component. Dworkin and MacKinnon responded to this problem during the public hearings in Minneapolis. In order to demonstrate the harms of porn, Dworkin and MacKinnon did not want to restrict the message of it but to allow those harmed by it to have the right to legal address. "Pornography is not what pornography says. If it were, the Ordinance's definition of pornography would be itself pornography, because it says exactly what pornography is. In other words, the Ordinance does not restrict pornography on the basis of its message" (1988: 37).

Moreover, they pointed to the way in which porn acts as an instance of the wider culture, but then contributes to it "The same message of sexualized misogyny pervades the culture—indeed, it does so more and more because pornography exists" (ibid.). Their analysis pointed to the way in which porn was disseminated into wider culture through marketing, advertising and dramatisations: writing about the abused permeated an entire cultural logic built on sex inequality of male dominance and female subordination. The Ordinances were repealed shortly after their introduction. One was a lawsuit instigated by a media group against the City of Indianapolis in 1985. They claimed it violated the First Amendment right of freedom of speech. Dworkin and MacKinnon did not classify the ordinance within the obscenity laws (moral regulation), but harm, they argued that pornographers were not just expressing ideas, but violating, torturing and harming women to do it. They wrote:

> So far as the Ordinance is concerned, all the ideas pornography expresses can be expressed—so long as coercion, force, assault, or trafficking in subordination is not involved. These are acts, not viewpoints or ideas. (58)

The tirade against Dworkin and MacKinnon resulted in the full mobilisation of the law to protect pornography as 'free speech.' They pointed out the contradiction in knowing that other forms of 'speech' can cause harm using the example of child abuse imagery (59) or images of lynchings of Black people in the American South—their purpose is to express a child rapist point of view, or the white supremacist point of view by doing it (ibid.).

The remainder of this section will present an overview of what porn is today, and explore the content of the 'speech' in it.

In 2020, Pornhub, one of the biggest providers of pornography, removed the majority of its content in response to an investigation. Over 13 million videos were removed, leaving 3 million as a result of a New York Times exposé (2020) that it featured child abuse content, the rape of trafficking victims and domestic violence. In 2015, Pornhub had 21.2 billion visits (Pornhub 2015) with 'teen,' and 'step sister' among its most popular categories. In another study, it was reported that 98 per cent of men and 73 per cent of women watched porn, but when asked about weekly use, the figure for women dropped to 26 per cent while remaining more or less stable for men at 80 per cent (Daspe et al. 2018). As Delicado-Moratalla (this volume) explains, the distinction between the commercial pornographer (who may use sets, script a 'narrative' and use paid prostituted women or men) and amateur pornographer is increasingly diminishing. This is reflected in the global production of porn known as 'amateur' or 'home-made' and can take place in private and personal settings or be illegally generated.

Pro-women feminists Dworkin and Mackinnon aimed to stop the mass normalisation of pornography in society, and they pointed to its brutalising practices by allowing those harmed by it the right to legally demonstrate harm. While their predictions on the realities of porn, the use of it in sexual relationships have proven to be prophetic—it has also sparked an unanticipated side-effect with the rise of new technologies. A whole raft of technologically enabled sex crimes now feature on pornographic websites, from child abuse, voyeurism, date-rape, spycam porn, revenge porn, internet facilitated sexual offending and up-skirting (Seto 2015).

Sexist Ethics Washing

In this section, I want to argue for why sex matters when making a case against porn robots, and why academic support for them does not bode well for how the field reflects on women and girls. The last two decades have witnessed a social experiment like no other: a generation in which men and boys (who vastly outnumber women and girls as users of porn) are engaging with sexuality, not as an interpersonal experience, but egocentrically mediated via screens which present others (women, girls, other men, boys) as anonymous third person(s). Could this depersonalisation of sexuality account for the cultural interest in porn robots?

The term sex is an umbrella term that clusters together several categories of meaning.

One of its meanings refers to sex as in sexed bodies (male or female) determined by reproductive, hormonal and genetic sex meanings that are under attack from transgender and queer theorists (Dansky 2021; Butler 2011). Made-up objects, of any kind, do not possess a biological sex.

Another way of understanding sex is as an abbreviation for sexual intercourse: an act that requires another who is participating. The word *intercourse* refers to "communication or dealings between individuals" (OED 1995: 709) and attention should be focused on the *inter* (intermediate, between, mutual, reciprocal) (ibid.: 708). Porn robots do not participate in any intercourse with buyers/users. These products are designed to look lifelike, and efforts are afoot to create more believable puppets of them—adding voice (aided by apps), movement (strategically placed motors) or heat (using synthetic cooling or heating devices).

Then there is masturbation which does not require another and is carried out without a participant. Surprisingly Levy, who in his 2007 book popularised the term 'sex robots' did not refer to them as masturbation robots despite focusing heavily on their use by 'lonely' men. Levy makes analogous arguments between what people do with each other. For example he draws on literature of the infant-mother developmental relationship and adult pair bonded attachments, and what (some) men do with products, for example, falling in love with their cars (2007).

There is something deliberate about the misuse of the term 'sex' that is associated with robots and dolls. There is clearly something *sexual* but it cannot be understood as analogous with relationship between adults participating in mutual sexuality.

There are political motivations to collapse the meaning of sex to include *everything* 'Once the pornographic is synonymous with the sexually arousing, anything that is sexually arousing might be pornographic' wrote Dworkin and MacKinnon (1988: 37) to explain that once sex and porn were indistinguishable, pornography would shape the culture more systematically. Moreover, demanding that mutuality be the hallmark of sexuality invokes a commitment to egalitarianism that invites a new way of thinking about love and sex among adults. A politics of love that prioritises relationship as connection and mutuality outside of property relations. This is the basis for the politics of love that is central to my work and the work of the campaign. It is not formed from 'writing about whores' but from living in connection with each other and calling for an end to property relations. People are not property, nor is property capable of anything other than fictional personification.

Links between the male-dominated technology industry and pornography are well documented. From Jimmy Wales, co-founder of Wikipedia who acquired his initial wealth from online pornography (Bomis.com); VHS won over Betamax because porn users preferred the format, streaming was developed and popularised on pornography sites before its mainstreaming application on platforms such as YouTube and Vimeo; and over 90 per cent of deep fakes are pornographic and not political (Gluck 2021). In the 1970s, a pornographic image from Playboy was used by Alexander Sawchuck at the University of Southern California Signal and Image Processing Institute—the model's name was Lenna (Hutchinson 2001). This was not the first time pornographic images of women were used. In the 1960s, images of Teddi Smith, a so-called Playboy 'playmate' was used by Lawrence G. Roberts in the 1960s for his Massachusetts Institute of Technology (MIT) master's thesis on image dithering (Roberts 1962). Why did Sawchuk and Roberts use these images? Because they were users of porn and brought it into their research labs. In 2012, Lenna's image was again used by a research group at the Agency for Science, Technology and Research (A*STAR) in Singapore (*BBC* 2012), this time using nanotechnology. The links between technology and porn continue unabated, but critical perspectives that problematise their intersections are few and far between. Some even going so far as to argue that pornography drives technology, as if human inventiveness rests on 'writing about whores.' If male technologists are to be believed, and innovation rests on the abuse of women and children, serious questions needs to be asked about its organisation, structure and function in the world.

In response to the growth of Big Tech that has taken place in the last twenty years, there has been a rise in the number of academics who position themselves as 'ethicists' recruited on prestigious and well-funded European and North American research projects, working for companies, or fronting organisations that deal directly with the ethics of robots and AI (Sharkey et al. 2017; Sandberg 2015; Richardson 2015). Some academics who have publicly called for bans in different cases such as autonomous weapons (see Sandberg 2015) have produced rationalisations for porn robots. As a matter of prudence, this begs the question to what extent a predominantly male, and pro Big Tech community can adequately represent the interests of women and girls in this area? When female academics have contributed to these issues, they focus on the explicit impacts of sexual objectification, while leaving the structural analysis of male power aside (Devlin 2018). Or when 'feminism' is 'queer' or 'cyborg' it flattens distinctions between humans and property (Haraway 2003, Kubes 2019) and subquently frames these RTH as progressive for women. Sometimes, academic 'queer' feminist approaches that frame pornography and prostitution as 'sex-work' are used by male academics against women to argue in favour of porn robots (Danaher 2019). Radical feminist perspectives meet with the same challenges in the 2000s as they did in the 1960s, 1970s and 1980s when they described what porn did, and how it did it.

At this stage arguably more porn robots exist in *academic texts* (Kubes 2019; Scheutz and Arnold 2016; Danaher 2019; Devlin 2018; Sharkey et al. 2017) than as discrete commercial products—there is apparently only one commercial porn robot for sale on the market—'Harmony'. Moreover, the bar for what turns a doll into a robot is set so low that anything that can aid its mimicry is credited as a robot. It is engineering and computating that turns these dolls into robots, away from fictional ideas into 'real' things, but they are merely further advances in ventriloquism and puppetry. Technologists produce many extraordinary and wonderful innovations, but they cannot transfer nor make humans (or animals) into made-up property forms. The expensively developed anthropomorphic robots are one step along the road of toys, not humans. Are academic narratives describing these effects? On the contrary, 'ethicists' are embellishing RTH with their own surplus of fictions and providing the rationalisations for their development. They are arguing for their dissemination as products in society for use in care homes (Fosch-Villarong and Poulsen 2020; Bendel 2015), or on prescription for paedophiles (Appel 2022). One might even compare the work of ethics used in the justification of porn robots to be forms of 'ethics washing.' Ethics washing is a term that has acquired more importance, particularly as there are more roles created for academics to work on technology projects and contribute to policy developments. Ethics washing occurs when problematic practices are recast as neutral or harmless—often justified in overly abstract philosophical terms. One example is a paper that invoked the work of Immanuel Kant to explore the use of child sex abuse robots for child sex predators (Gordon and Nyholm 2022). There has to be some legitimate basis to ethics otherwise it becomes Aristotle's anti-woman, children and pro-slavery politics. As a guiding principle ethics must take into account the humanity of women and children.

From a radical feminist point of view, the road to porn robots will be paved by the success of porn dolls in altering men's sexual tastes and transforming women's sexual boundaries as can happen through pornographic reshaping of sex, and sexual norms (Dines 2010). Hence the passing off of porn dolls as robots is central to the enterprise of paving the ground for the latter's development. A market, catering for tastes in plastic, requires cultivation. Moreover, approximated versions of pornography in the form of dolls and/or robots are expensive to purchase (from £5000 for dolls for some options). The curiosity, fuelled by a 'don't worry' ethics community forms the backdrop to their experimentation among male users who can now test them out by renting them at 'brothels.' Porn robots, like porn on the internet has spilled out into the lives of women, and there are fewer spaces where women and girls can be without pornography intruding in on their lives. Men's porn is protected, and elaborated porn dolls and robots are washed ethically by a professional academic establishment. The presence of porn robots in public spaces (museums, wheeled around in supermarkets, taken on the school run, put at the breakfast table with children, put in art exhibitions, on TV shows and in films) represents the triumph of men to display their pornography in public, and force women and children to participate in their sexual fetishes (see Jeffreys this volume), while women and girls witness their own dehumanisation and depersonalisation before their eyes.

Conclusion: The Politics of Love

A radical feminist lens of technology and pornography is urgently required to make sense of new forms of porn and account for the harms that produce it and are reinforced by its practices.

This book will be a significant intervention in how 'sex' robots as specific examples of RTH of women and girls are understood, interpreted and made sense of leading to huge potential to cross disciplinary boundaries and engage researchers in the social sciences, arts, humanities and technology. The work of the Campaign Against Porn Robots will continue to spell out the harms of these technologies for women and girls. Men and boys are also psychically wounded in the process, as their capacities for love, intimacy, and connection are also destroyed when they align with sexist practices and support, use and produce pornography. While porn robots are not yet mass produced, there is an urgent need for research to examine how men's use of porn dolls is impacting on women and children who live in close proximity to users of them, as this will give us an insight into how a porn robot industry will damage human relationships. Therefore, this edited volume is so crucial to those who want to rail against this public assault on women's humanity, its vicious dehumanisation. We want technologies that provide opportunities for women and girls to be written about as human: intelligent, capable, imaginative, visionary, strong, revolutionary and transformative. These are the writings of women and girls on the surfaces of RTH we want to create, but they will not exist in technology while women and girls are regarded as 'whores' rather than as human and the *writing of her abuse* will continue.

References

Appel, J.M. 2022. Unconventional harm reduction interventions for minor-attracted persons. *Clinical Ethics*. https://doi.org/10.1177/14777509221117981.

Armer, P. 1963. Attitudes towards intelligent machines. In *Computers and thought*, eds. E. Feigenbaun and J. Feldman. New York and San Francisco: McGraw-Hill book Company.

BBC. 2012. *Playboy Centrefold Photo Shrunk to Width of Human Hair*. 14 August 2012. Accessed 21 Nov 2022. https://www.bbc.co.uk/news/technology-19260550

Bendel, O. 2015. Surgical, therapeutic, nursing and sex robots in machine and information ethics. In *Machine medical ethics*. Cham: Springer, 17–32.

Bindel, J. 2019a. The horror of big porn: Why pornography apologists are wrong. *The Spectator*, 1 October 2019 [Online]. https://spectatorworld.com/topic/horror-big-porn/. Accessed 15 Aug 2022.

Bindel, J. 2019b. *The Pimping of Prostitution*. London: Palgrave Macmillan.

Benjamin, Walter. 1935 (1969). *The Work of Art in the Age of Mechanical Reproduction*. In (Arendt, H.). Illuminations. Translated by Harry Zohn. NY: Schocken Books.

Benjamin, R. 2019. *Race after technology: Abolitionist tools for the new Jim code*. Cambridge: Polity Press.

Butler, Judith. 2011. *Bodies that Matter: On the Discursive Limits of Sex*. London and New York: Routledge.

Campaign Against Porn Robots, https://campaignagainstsexrobots.org/.

Campaign to Stop Killer Robots, https://www.stopkillerrobots.org/.

Côté, J.E., and C.G. Levine. 2014. *Identity, formation, agency, and culture: A social psychological synthesis*. Psychology Press.

Culture Reframed: Building Resilience and Resistance to Hypersexualised media and porn. http://culturereframed.com/.

Danaher, J. 2019. Building better sex robots: Lessons from feminist pornography. In *AI love you*. Cham: Springer, 133–147.

Dansky, Kara. 2021. *The abolition of sex: How the transgender agenda harms women and girls*. New York and Nashville: Bombardier Books.

Daspe, M.È., Vaillancourt-Morel, M.P., Lussier, Y., Sabourin, S. and Ferron, A., 2018. When pornography use feels out of control: The moderation effect of relationship and sexual satisfaction. *Journal of Sex & Marital Therapy*, 44 (4), pp. 343–353.

Devlin, K. 2018. *Turned on: Science, sex and robots*. London: Bloomsbury Publishing.

Dines, G. 2010. *Pornland: How porn has hijacked our sexuality*, Boston, MA: Beacon.

Dworkin, A. 1988. *Intercourse*. London: Arrow.

Dworkin, A. 1989. *Pornography: Men possessing women*. London: Women's Press.

Dworkin, A., and C. MacKinnon. 1988. *Pornography and civil rights: A new day for women's equality. Organizing against pornography*. London: Women's Press.

Erich, Auerbach. 1953. *Mimesis: The Representation of Reality in Western Literature*. Translated by Willard R. Trask. Princeton University Press: Princeton.

Feminist Ethical Dialogues. 2020. Campaign against porn robots [Online]. *Youtube*. https://www.youtube.com/watch?v=kbVTMGnHVQU&list=PLHTHH0NwXeXPmGoyTosG-3ykeGdN5A6Co.

Fosch-Villaronga, E., and A. Poulsen. 2020. Sex care robots. *Paladyn, Journal of Behavioral Robotics* 11 (1): 1–18.

Fritz, N., V. Malic, B. Paul, and Y. Zhou. 2020. A descriptive analysis of the types, targets, and relative frequency of aggression in mainstream pornography. *Archives of Sexual Behavior* 49 (8): 3041–3053.

Gluck, G. 2021. Deepfakes and VR porn. With Kathleen Richardson. Feminist ethical dialogues [Online]. *Youtube*. Accessed 8 Jan 2022. https://www.youtube.com/watch?v=oQj0sFI9K7E&t=3542s.

Gordon, J.S., and S. Nyholm. 2022. Kantianism and the problem of child sex robots. *Journal of Applied Philosophy* 39 (1): 132–147.

Harper, C.A., and R. Lievesley. 2020. Sex doll ownership: An agenda for research. *Current Psychiatry Reports* 22 (10): 1–8.

Haraway, Donna. J., 2003. *The Companion Species Manifesto: Dogs, People, and Significant Otherness* (Vol. 1, pp. 3–17). Chicago: Prickly Paradigm Press.

Hicks, Marie. 2017. *Programmed Inequality: How Britain Discarded Women Technologists and Lost its Edge In Computing*. Cambridge: MIT Press.

Hooks, B. 2000. *Feminist theory: From margin to center*. London: Pluto Press.

Hutchinson, Jamie. 2001. Culture, Communication, and an Information Age Madonna. IEEE Professional Communication Society Newsletter. May/June 2001. 45 (3). pp. 1–7

Jecker, N.S. 2021. Nothing to be ashamed of: Sex robots for older adults with disabilities. *Journal of Medical Ethics* 47 (1): 26–32.

Jeffreys, Sheila. 1990. *Anticlimax a Feminist Perspective on the Sexual Revolution*. The English Women's Press. Ltd.

Jeffreys, S. 2005. *Beauty and misogyny: Harmful cultural practices in the West*. Hove: Routledge.

Kaminer, W. 2000. Forward to the new edition. In *Defending pornography: Free speech, sex, and the fight for women's rights*, ed. N. Strossen. New York: NYU Press.

Karaian, L. 2022. Plastic fantastic: Sex robots and/as sexual fantasy. *Sexualities*. https://doi.org/10.1177/13634607221106667.

Keuls, E.C. 1993. *The reign of the phallus: sexual politics in ancient Athens*. Berkeley: University of California Press.

Kubes, T. 2019. New materialist perspectives on sex robots. A feminist dystopia/utopia? *Social Sciences*, 8 (8), P. 224.

Richardson, Kathleen. 2016. Sex Robot Matters: Slavery, the Prostituted, and the Rights of Machines. *IEEE Technology and Society Magazine*, 35 (2) pp. 46–53.

Latour, B. 2012. *We have never been modern*. Cambridge: Harvard University Press.

Lerner, Gerda. 1986. *The creation of patriarchy (Vol. 1). Women and History*; V. 1. Oxford: Oxford University Press.

Levy, D. 2007. *Love and sex with robots: The evolution of human-robot relationships*. Wiltshire: Duckworth, 352.

MacKinnon, C.A. 1993. *Only words*. Cambridge, Mass: Harvard University Press.

McNally, L. 2015. Business as usual, rebranded as ethics: the whitewashing of systemic injustice. In *Freedom fallacy: The limits of liberal feminism*. Ballarat: Connor Court Publishing, 105–112.

Michael, Taussig. 1993. *Mimesis and Alterity: A Particular History of the Senses.* London and New York: Routledge.

Minsky, M. 1967. *Computation: Finite and infinite machines*. Englewood Cliffs, NJ: Prentice-Hall Inc.

Nyholm, Sven, and E. Frank Lily. 2017. From sex robots to love robots: Is mutual love with a robot possible? In *Robot sex: Social and ethical implications*, eds. John Danaher and Neil McArthur. Cambridge: MIT Press.

OED. 1995. *The concise Oxford dictionary of current English*, 9th ed., ed. Della Thompson. Oxford: Clarendon Press.

Peeters, A., and P. Haselager. 2021. Designing virtuous sex robots. *International Journal of Social Robotics* 13 (1): 55–66.

Pornhub. 2015. Year in review. https://www.pornhub.com/insights/pornhub-2015-year-in-review. Accessed 5 Aug 2022.

Prostasia, Does Congress care more about dolls than real children? Protasia Foundation. Accessed 25 Aug 2022.

Protasia Foundation. 2022. Does fantasy cause or prevent offending? [Online] https://prostasia.org/project/research-fund/. Accessed 25 Aug 2022.

Prostasia Foundation. n.d. Does Congress care more about dolls than real children? https://prostasia.org/campaign/does-congress-care-more-about-dolls-than-real-children/. Accessed 21 Nov 2022.

Richardson, K. 2015. *An anthropology of robots and AI: Annihilation anxiety and machines*. London and New York: Routledge.

Richardson, K., 2016. Sex robot matters: slavery, the prostituted, and the rights of machines. *IEEE Technology and Society Magazine*, 35 (2), pp.46-53.

Roberts, G. L. 1962. Picture coding using pseudo-random noise. *IRE Transactions on Information Theory*, 8 (2). pp. 145–154

Richardson, K. 2018. *Challenging sociality: An anthropology of robots, autism, and attachment*. New York: Springer.

Sandberg. A. 2015. Sex and death among the robots: When should we campaign to ban robots? http://blog.practicalethics.ox.ac.uk/2015/09/sex-and-death-among-the-robots-when-should-we-campaign-to-ban-robots/. Accessed 5 Aug 2022.

Scheutz, M., and T. Arnold. 2016. Are we ready for sex robots? In *2016 11th ACM/IEEE International Conference on Human-Robot Interaction (HRI)*. Christchurch: IEEE, 351–358.

Seto, M.C. 2015. *Internet-facilitated sexual offending*. Washington, DC: US Department of Justice, Office of Justice Programs, Office of Sex Offender Sentencing, Monitoring, Apprehending, Registering, and Tracking.

Sharkey, N., A. van Wynsberghe, S. Robbins, and E. Hancock 2017. *Our sexual future with robots*. The Hague: Foundation for Responsible Robotics. http://responsiblerobotics.org/wp-content/uploads/2017/07/FRR-Consultation-Report-Our-Sexual-Future-with-robots_Final.pdf.

Smith, M., 2013. The erotic doll: a modern fetish. New Haven and London: Yale University Press.

Sparrow, R. 2017. Robots, rape, and representation. *International Journal of Social Robotics* 9 (4): 465–477.

Stock, K. 2021. *Material girls: Why reality matters for feminism*. UK: Hachette.

Strossen, N. 2000. *Defending pornography: Free speech, sex, and the fight for women's rights*. New York: NYU Press.

Vaillancourt, M.P., M.È. Daspe, V. Charbonneau-Lefebvre, M. Bosisio, and S. Bergeron. 2019. Pornography use in adult mixed-sex romantic relationships: Context and correlates. *Current Sexual Health Reports* 11 (1): 35–43.

Wallin, D.J. 2007. *Attachment in psychotherapy*. New York: Guilford press.

Yeung, K., A. Howes, and G. Pogrebna. 2019. AI governance by human rights-centred design, deliberation and oversight: An end to ethics washing. In *The Oxford handbook of AI ethics*. Oxford: Oxford University Press.

Correction to: Man-Made Women

Kathleen Richardson and Charlotta Odlind

Correction to:
K. Richardson and C. Odlind (eds.), *Man-Made Women*, Social and Cultural Studies of Robots and AI, https://doi.org/10.1007/978-3-031-19381-1

The original version of Chapters 1 and 2 were inadvertently published without Abstract in the online version. The online version of the chapters have been updated with Abstract now, and can be viewed on Chapters 1 and 2.

The updated original versions of these chapters can be found at
https://doi.org/10.1007/978-3-031-19381-1_1
https://doi.org/10.1007/978-3-031-19381-1_2

© The Author(s), under exclusive license to Springer Nature Switzerland AG 2023

K. Richardson and C. Odlind (eds.), *Man-Made Women*,
Social and Cultural Studies of Robots and AI,
https://doi.org/10.1007/978-3-031-19381-1_10

Index

A

Abolitionism, 11, 94
Acton, William, 146
Age of consent laws, 119–121
Animation, reanimation, bringing 'to life', 57, 152, 175, 178
Art
- artists, 11, 17–19, 24, 69–72, 74, 75, 79, 83–85, 142, 143, 147, 152, 155, 157, 166
- avant-garde, 72, 157
- gallery, 83, 126, 152
- history, 2, 11, 19, 24, 53, 74, 103, 105, 119, 139, 143, 165
- theory, 9, 12, 18, 20, 27, 74, 79

Artificial Intelligence (AI), 4, 5, 56, 83, 85, 93, 98–100, 103–108, 118, 140, 141, 163, 173, 174, 179, 186
Audio pornography, 156, 162
Australian Institute of Criminology, 128

B

Balesteri, Frank, 157
Banks, Grace, 84
Barry, Kathleen, 78, 82, 83, 85, 86, 96, 152
Barthes, Roland, 71, 160
BDSM, 74, 82, 125, 160
Beatrice, 142, 144, 151, 160
Beauty standards, 82
Beauvoir, Simone de, 21, 71, 93, 142, 144
Bellmer, Hans, 69–76, 86
Benjamin, Walter, 173, 176
Bijoux Indiscrets, 161
'Blue discs'/party records, 154
Bondage, 97
- jewellery, 161
- manacles, 75

Breath, 'breath play', 142, 14[illegible], 154, 157, 162, 164
Bronfen, Elizabeth, 79
Brothel, 38–41, 44, 73, 152, 153, 187

© The Editor(s) (if applicable) and The Author(s), under exclusive license to Springer Nature Switzerland AG 2022

K. Richardson and C. Odlind (eds.), *Man-Made Women*, Social and Cultural Studies of Robots and AI, https://doi.org/10.1007/978-3-031-19381-1

Brumfit, Stuart, 75, 76

C

Califia, Pat, 118, 121, 122
Campaign Against Sex Robots/Campaign Against Porn Robots, 5, 11, 172
Cantor, James, 118, 123–125
Carry on Screaming, 10, 67, 68
Child abuse, 5–8, 123, 126, 130, 178, 183
Child sex abuse dolls, made in the likeness of actual children, 130
Child sexual abuse, 12, 52, 58, 59, 61, 118–120, 123, 124, 126–129, 132, 178
Child sexual abuse material, 118, 121, 124, 127, 129, 130
Child sexuality, 119, 120
Choice, 2, 26, 82, 83, 123
Choking, 165
Consent, 59, 74, 78, 81–83, 85, 123, 160, 177
Corbett, John, 155
Cosmetic surgery, 8, 25, 26
Cummings, Whitney, 163, 164
Cyborg, 100, 179, 186

D

Daly, Mary, 79, 86
Danaher, John, 3, 36, 37, 61, 104, 186
Dante, 142
Dead body, 68, 70, 76–78, 94, 145
Decadent art and artists, 140, 143, 145
Deepfakes, 151, 173
Dehumanisation, 5, 7, 25, 77, 78, 173, 187, 188
De Palma, Laura, 75, 76
Destigmatising paedophilia, 123
Devlin, Kate, 3, 5, 7, 18, 96, 186
Dines, Gail, 6, 22, 23, 127, 128, 172, 180, 181, 187
Dissociation, 83, 84, 86, 162, 164
Doll brothel, 3, 36, 39–42, 152
Doll forum, 22, 80
Doll rape, 42
Domestic violence/abuse, 85, 146, 165, 183
Dominance/subordination, 7, 19, 27, 43, 52, 92, 94, 97, 101, 108, 129, 182
Dorfman, Elena, 84
Döring, Mia, 153
Dramatisations, 173, 175
Drawing, 69–73, 102, 125, 127, 129, 175
Dworkin, Andrea, 36, 42, 50, 52, 53, 69, 71, 72, 78, 79, 81, 86, 94, 100, 172, 175, 180–183, 185

E

Eckstein, Emma, 148, 149, 166
Edison, Thomas, 143, 151
Ekman, Kajsa Ekis, 140, 152, 153
Ethics, 2–5, 12, 80, 107, 172, 179, 186, 187
Ethics washing, 172, 184, 187
Eulogy, 7, 160

F

Fantasy, 8, 12, 23, 94, 95, 99, 101, 102, 105, 106, 126, 128, 131, 149, 155, 177, 179
Fantasy, psychical, 148
Fashion, 10, 21, 68, 75, 76, 80, 82, 152, 165
Female pleasure, 155, 161
Femininity, 18, 19, 24, 26, 41, 146, 150, 160
Feminism

first wave, 145
liberal, 26, 82, 164, 173
radical, 3, 4, 6, 9, 50, 52, 95, 119, 140, 145, 146, 153, 172, 175, 180, 186–188
second wave, 9, 119
Feminist campaigns, 104
Feminist theory, 35, 78, 79
Fetish, 3, 7, 9, 12, 23, 24, 49, 50, 52, 53, 56, 60, 61, 95, 103–105, 165
Fetishism, 24, 50, 52, 55–58, 60, 61, 103, 164
Fiction, 61, 102, 140, 141, 179
Film, 10, 18, 19, 25, 27, 43, 51, 67, 68, 73, 74, 76, 78, 80, 81, 98, 140, 151, 152, 155, 161, 174–176, 180, 187
Fliess, Wilhelm, 148
Frankenstein, 68
Frank, Lily Eva, 37, 141, 179
Free speech, 6, 172, 180, 183
Freud, Sigmund, 11, 52–54, 56, 119, 148–150, 166
Fuller, David, 76, 77

G

Gender, sex/gender, 10, 20, 24, 25, 27, 44, 52, 53, 72, 95, 100, 104, 107
'Girlfriend' experience, 85, 153
Goertzel, Ben, 80

H

Harmony, 37, 38, 99, 102, 103, 106, 186
Harmony Show, The, 84, 85
Hawk Swanson, Amber, 84, 85
Hennessy, Don, 165
Henningsen, Therese, 152
Herman, Judith, 26, 119, 148
Hershman Leeson, Lynn, 84
Humour, sexist jokes, comedy, 10, 68, 154
Huysmans, J-K, 162, 163, 166
Hysteria, 146

I

iDollators, 84
Imitation, 41, 44, 92, 95, 98, 104, 158, 173
Incels, 95
Incest, 119, 129
Instinct, 55, 150, 165
Internet, 24, 36, 39, 43, 51, 98, 127, 177
Irigaray, Luce, 79

J

Jeffreys, Sheila, 6, 9, 19, 23, 24, 36, 42, 44, 50–55, 58, 59, 68, 76, 82, 83, 86, 96, 102, 119, 140, 143, 145–147, 157, 158, 160, 165, 180, 181, 187

K

Kant, 74, 152, 172, 187
Kappeler, Susanne, 75, 78, 79, 81, 86, 139, 140, 151, 152, 158, 159, 161, 164, 166
Kapsalis, Terri, 155
Kasire, Thomas, 81
Keuls, Eva, 176
Kinetoscope/Mutoscope, 151
Kink, 'kink-shaming', 74, 82, 125, 177
Kinsey, Alfred, 119
Knight, Nick, 75

L

Laws/laws criminalising child sex abuse dolls, 8, 20, 51, 58, 59, 68, 79, 81, 93, 118–123, 125, 131, 177, 181–183
Leane, Shaun, 75
Leigh, Stacy, 84
Lenna, 185
Lerner, Gerda, 19–21, 27, 97, 173
Levy, David, 3, 36, 37, 79, 80, 96, 100, 124, 174
Liberation, 1, 6, 57–59, 72, 82
Lichtenstein, Therese, 71, 72
Literature, 10, 23, 49, 50, 54, 59, 68, 79, 86, 107, 140–142, 147, 172
Love
  and death, 143, 144, 160
  interpersonal mutuality, 166
  philosophical conceptualisations of, 80
  sovereignty/abandonment, 144, 160
  with robot, 3
'Love dolls', 22, 37, 141
Lying, 'hysterical mendacity', 73, 126, 145, 149, 150

M

MacKinnon, Catharine A., 6, 50, 52, 78, 79, 83, 93, 172, 180–183, 185
Maintenancer, 152, 153
Male sex right, 7, 11, 50, 53, 96, 128, 132, 158, 162
Mannequin, 4, 23, 68, 86, 130
Marriage
  arranged, 51, 141
  to robot, 37, 80, 141
Masculinity, 43, 52
Masson, Jeffrey, 148, 149
Masturbation, 37, 61, 94, 126, 149, 151, 164, 184
McKay, Anne, 153
McMullen, Matt, 37, 79, 80, 83, 85, 86
McQueen, Alexander, 10, 75, 76
Mechanisation of sexuality, 9, 44
Meineche Hanson, Sidsel, 152
Mimesis, 173, 174
Mimicry, 173, 175, 186
Minneapolis Ordinance, 181
Minor Attracted Persons (MAPs), 10, 123, 178
Misogyny, 4, 76, 94, 163, 165, 182
Mitrani, Nora, 73
Moaning in music, 140, 155
'Moaning'-sexual harassment, 158
Moral panic, 118, 121
Moran, Rachel, 6, 101, 102, 153
Mythology, 18, 72, 144, 174
  Greek, 105

N

Nead, Lynda, 74
Necrophilia, 23, 69, 76–78
Needham, Gerald, 143
Newman, Michael, 72, 74
Nyholm, Sven, 37, 141, 172, 179

O

Objectification, 6, 10, 12, 18, 23, 26, 36, 41, 43, 68, 78, 79, 94, 103, 107, 127, 146, 147, 155, 172, 179, 186
'Obscene' phone calls, 157
Onlyfans, 151
Ontology, 174
Oppression, 1, 9, 21, 50, 52, 53, 78, 92, 93, 101, 148
Orgasms, Library of, 160, 162
Orgasm sound, 155, 156, 160
Ovid, 17–20, 23, 24, 27, 105

P

Paedophiles/paedophilia, 3, 10, 13, 52, 59, 72, 104, 119–125
Paedophiles as victims, 123
Pankhurst, Christabel, 146, 147
Paraphilias, 9, 52, 54, 57–60, 62, 82
Parent-Duchâtelet, Alexandre, 153
Patriarchal societies, 35, 79
Peep-show, 11, 139–141, 143, 145, 147, 149, 151, 153, 155, 157, 159, 161, 163–167, 169
Phonographs, of women's voices, 153
Photographs, photography, 69–74, 76, 80, 81, 84, 122, 142, 143, 175, 176
Playboy, 180, 185
Political speech, women's, 140
Pop music, 156
Popular culture, 3, 10, 83, 147
Porn, 5, 6, 11, 22, 36, 38, 39, 41, 43, 81, 82, 92, 104, 109, 122, 127, 151, 153, 154, 156, 158, 165, 172, 175, 176, 180–183, 185, 188
Porn dolls, 7, 11–13, 17, 18, 20–24, 27, 36, 43, 94, 96, 99–104, 107, 171, 172, 176, 187, 188
pórnē, 175
Pornhub, 38, 42, 81, 129, 183
Pornography
  ancient, 144
  DIY, 80
  doll, 18, 38, 39, 43, 44, 60, 61, 80, 84, 140
  'home-made', 183
  industrial, 143
  internet, 39, 187
  nineteenth century, 54
  snuff, 71
Porn robot(s), 2, 8–13, 27, 36, 44, 92–108, 172–174, 176, 177, 179, 180, 184, 186, 187
Pose, the pose for the photograph, 81, 86, 92
Poupee, La, 75
Prostasia Foundation, 118, 124, 125, 127
Prostitution, 5–9, 11, 12 39, 40, 50, 61, 69, 76, 78, 80, 82–86, 94–96, 100–102, 127, 142, 143, 146, 152, 153, 160, 166, 172
Psychephilia, 85, 86
Psychoanalysis, 11, 119, 140, 147, 148, 150, 165, 166
Public spaces, 4, 68, 151, 155, 187
Pygmalion, 8, 17–20, 23, 24, 27, 57, 104–106

Q

Queer theory, 121

R

Radical feminism, 95
Radio, 155–157, 160, 164
'Realbotix', 37, 85, 140
Realdolls, 84, 141, 164
Recording, sound, 161, 180
Records, phonographic, 73, 81, 122, 153, 156, 161
Reeves, Joyce, 72
Representation, 26, 41, 78, 79, 81, 86, 153, 155, 156, 173, 174, 176
Representation, structure of, 79, 81
Rexer, Raisa, 143
Richardson, Kathleen, 2, 4, 5, 8, 11, 36, 92, 100, 107, 108, 140, 173, 177, 186
'Robosexual', 3, 84, 99
Roper, Caitlin, 10, 104, 126, 129, 131, 158, 178
Rubin, Gayle, 79, 118, 120–122

S

Sade, Marquis de, 10, 71, 72, 81, 159, 160
Sadomasochism, 41, 42, 50, 57, 59, 71
Safeguarding, 8, 125
Schwartz, Hillel, 166
Schwob, Marcel, 142–145, 166
Sculpture, 18, 69, 144, 160
Sex doll art, 83
Sex doll(s), 2–4, 7–9, 12, 18, 19, 22, 26, 36–39, 41–44, 49, 56, 58, 60, 61, 68–70, 76, 77, 79–81, 83–86, 101, 104, 117, 124, 126, 130, 131, 139–142, 151–153, 164, 166, 177, 179
Sex industry, 11, 69, 70, 78, 83, 92, 96, 103, 142, 151, 157, 164, 166
Sexology, 51, 53, 54, 59, 82, 147, 150, 161
Sex robot, 2–9, 11, 12, 36–40, 43, 44, 49, 56, 68, 85, 86, 92–94, 102, 140–142, 153, 156, 160, 163, 164, 174, 179, 184
Sex, sex/gender, 2, 3, 6, 10, 12, 13, 20, 23, 27, 37, 38, 40, 41, 49–51, 53–55, 57, 59, 61, 68, 73, 74, 77, 78, 81–85, 93–97, 100–102, 104, 106, 108, 117–120, 124–132, 143, 146, 147, 149, 151, 155–157, 160, 161, 164–166, 171–173, 175–182, 184, 185, 187
Sex shows/live sex shows, 151, 152
Sex/tech fetishisation, 156
Sex toys, 125, 126, 161
Sexual abuse, 11, 77, 82, 118–120, 123–129, 131, 132, 146, 148–150, 175, 178
Sexual assault/violence, 6, 39, 42, 51, 58, 71, 72, 77, 84, 102, 147, 150
Sexual exploitation, 6, 9, 12, 83, 127, 130, 147, 149
Sexual harassment, 50, 51, 146, 157, 158
Sexual intercourse, 40, 52, 184
Sexualisation of girls, 10
Sexual liberation, 58, 118–120, 125, 165, 181
Sexual Offences Act (British), 76
Sexual outlets, 10, 123–125, 132, 177
Sexual politics, 44, 84, 141, 145–147
Sexual repression, 72, 118, 120
Sexual revolution, 42, 57, 58, 82, 143, 147, 150, 160
Sexual rights, 95, 118, 121, 124
Shaw, Debra, 75
Showalter, Elaine, 144, 147, 153
Silencing women, 147, 158
Silent George, 154
'Silicone attraction', 84
Simmons, Laurie, 84
Spender, Dale, 145
Stigma, 119, 121, 152
Subject object relationships, 36, 79
Suffragettes, 145–147
Surplus of fiction, 174, 179, 186
Surrealism, 75
Syphilis, 144, 145, 150, 162

T

Technology
  computational, 176
  developments, 21, 118, 131, 132, 141, 143, 151, 176
  digital, 11, 155, 166, 176
  historical, 98, 118
  materials, 70, 108, 164
  representational, 173, 174

Teen porn, 129
Telephone sex lines, 156
Television, 40, 67, 68, 155, 174
Theatre, burlesque, 153
Thinking Sex, 121, 122
Torture, 22, 40–42, 44, 76, 78, 81, 129, 131, 156, 160, 177
Transgenderism, 100, 184
Transhumanism, 11, 162, 164, 166
Trottla, 126

U

Unconscious, theory of, 150

V

Ventriloquism, 11, 162, 163, 166
Victimisation, 77, 78, 81, 121
Victimless crime, 127
Victim, willing and unwilling, 86, 140, 159–161, 166
Video, 3, 11, 21, 22, 38–40, 80, 129, 130, 158, 166, 183
Virtual child sexual abuse material, 125, 127, 130
Voice, 11, 106, 140, 141, 143–145, 150, 153, 154, 157–159, 162–167, 172, 184
Voice, recordings
  political voice, 140, 166
  women's voices, 11, 140, 145, 147, 149, 153, 155, 158, 161, 162, 166

W

Webb, Peter, 70–72, 74, 79, 86
When Harry Met Sally, 142, 155
Whore, 144, 153
  adoption as everyday term, 152
  pórnē, 175
  writing about whores, 175, 185
Wilde, Oscar, 147
Wilson,Sarah, 73
Woman-sewer, 153, 164

Y

Youtube, 185

Z

Zürn, Unica, 69, 73, 74

Printed by Printforce, United Kingdom